A Practical Course in Advanced Structural Design

Tim Huff, P.E., Ph.D.

T0298318

CRC Press
Taylor & Francis Group
Boca Raton London New York

CRC Press is an imprint of the
Taylor & Francis Group, an **informa** business

First edition published 2021 by
CRC Press
6000 Broken Sound Parkway NW, Suite 300, Boca Raton, FL 33487-2742

and by
CRC Press
2 Park Square, Milton Park, Abingdon, Oxon, OX14 4RN

© 2021 Tim Huff
CRC Press is an imprint of Taylor & Francis Group, LLC

Library of Congress Cataloging-in-Publication Data

Names: Huff, Tim, author.
Title: A practical course in advanced structural design / Tim Huff, P.E.,
Ph.D.
Description: First edition. | Boca Raton, FL : CRC Press, 2021. | Includes
bibliographical references and index.
Identifiers: LCCN 2020046722 (print) | LCCN 2020046723 (ebook) | ISBN
9780367746667 (hardback) | ISBN 9781003158998 (ebook)
Subjects: LCSH: Structural design--Textbooks.
Classification: LCC TA658 .H84 2021 (print) | LCC TA658 (ebook) | DDC
624.1/771--dc23
LC record available at https://lccn.loc.gov/2020046722
LC ebook record available at https://lccn.loc.gov/2020046723

ISBN: 978-0-367-74666-7 (hbk)
ISBN: 978-0-367-74668-1 (pbk)
ISBN: 978-1-003-15899-8 (ebk)

Typeset in Times
by Deanta Global Publishing Services, Chennai, India

Access the eResource: www.routledge.com/9780367746667/SolutionsManual

Contents

Preface

This book is written from the perspective of a practicing engineer with 35 years of experience now working in the academic world to pass on lessons learned over the course of a structural engineering career. The topics covered in this book will enable the beginning structural engineer to gain an advanced understanding prior to entering the workforce.

The textbook will also be of use to practicing engineers, as the topics covered are based on theory but are encountered frequently in practice. Practical example problems are provided throughout the textbook.

Having worked in both building design and bridge design, as well as other areas of structural design, the author has attempted to include topics which may receive little or no attention in a typical undergraduate curriculum, but which beginning engineers are often asked to address early in their careers.

About the Author

Tim Huff has 35 years of experience as a practicing structural engineer. Huff has worked on building and bridge projects in the United States and has contributed to projects in India, Ethiopia, Brazil, the Philippines, and Haiti as a volunteer structural engineer with Engineering Ministries International. He is a faculty member of the Civil & Environmental Engineering Department at Tennessee Technological University in Cookeville, where he resides with his beautiful and talented wife, Monica, an artist and teacher.

Acknowledgments

For my family of students, I am grateful. You have taught me a lot. For the family of my childhood, I am blessed. Mom and Dad, Troy and Holli – I love you very much, and you are a huge part of the inspiration for all my endeavours. For Majo and Esteban – I find you to be exceptional, resilient, and full of joy in most any circumstance. I will keep working on that myself. Finally, I thank my beautiful, only wife, Monica, for your love and support and a nudge to get this done. Your art and teaching leave me speechless most of the time.

For the reader, I hope you will find this a useful tool, and . . .
"The LORD bless you and keep you;
The LORD make his face shine upon you,
And be gracious to you.
The LORD lift up his countenance upon you,
And give you Peace." – Numbers 6:24–26, New King James Bible

"What is hateful to you, do not do to your fellow: this is the whole Torah; the rest is the explanation; go and learn." – Hillel the Elder

"If you don't know where you're going, you'll end up somewhere else." – Yogi Berra

1 Introduction

This text is intended for a senior level course for civil engineering students whose focus is structural engineering. The material may also be useful for graduate level courses and as a reference for practicing structural engineers. First courses in structural analysis, structural steel design, and reinforced concrete design should be considered prerequisites to the material covered here.

The basis of this book is course notes developed by the author for courses in advanced structural design, bridge design, and earthquake engineering at Tennessee State University and Tennessee Technological University after 35 years of structural engineering practice.

The material in this book may be studied most effectively with several tools in hand. These are freely available.

1. AISC 360-16 Specification for Structural Steel Buildings
2. AISC 341-16 Seismic Provisions for Structural Steel Buildings
3. AISC 358-16 Prequalified Connections for Special and Intermediate Steel Moment Frames for Seismic Applications
4. SeismoStruct Educational Version – Nonlinear Static & Dynamic Structural Analysis
5. SeismoMatch Educational Version – Spectral Matching in the Time Domain
6. SeismoArtif Educational Version – Spectral Matching in the Frequency Domain
7. IES VisualAnalysis Educational Version – Structural Design Software
8. SigmaSpectra – Ground Motion Scaling and Selection Tool

The AISC Standards may be downloaded from aisc.org. The SeismoSoft applications are available from seismosoft.com. IES VisualAnalysis Educational can be downloaded from iesweb.com/edu. SigmaSpectra, by Albert Kottke, is available at GitHub.

The book is not intended to be a comprehensive treatment of any single subject in the field of structural engineering, but to familiarize and summarize for the structural engineering advanced student as well as the practicing engineer, a variety of topics.

While it is most often necessary to use software in modern structural design, the engineer should have the ability to perform sanity checks on software results, and to estimate values for design parameters using hand calculations and simplifications. The material presented here will be of assistance in completing such tasks.

Topics from both building and bridge design are included and serve to enhance an undergraduate curriculum. Examples are included throughout each chapter.

Chapter 2 presents a discussion of various types of structural analysis including linear elastic versus nonlinear analysis, first-order versus second-order analysis, and response spectrum analysis versus response history analysis. The chapter also includes a brief discussion on seismic site response analysis. Finally, a detailed presentation of the substitute structure method for inelastic seismic response is presented.

Chapter 3 presents topics unique to the design of buildings. Composite beam design is discussed, followed by an outline and examples of the AISC direct analysis method for stability. Plastic analysis techniques, important in seismic and blast-resistant design, are covered, as are requirements from design specifications related to plastic design, also known as inelastic design. Various lateral force resisting systems and design philosophies are covered prior to connection design in steel. Issues related to computer modeling of buildings, vertical and horizontal seismic load distribution, and the design of moment resistant column bases finish out Chapter 3.

Bridge loads, limit states, and load combinations are discussed in detail in Chapter 4, followed by a presentation of issues related to both prestressed concrete and structural steel superstructures for bridges. Substructures and foundation systems commonly used in bridges are covered in Chapter 4. Earthquake effects on bridges, in terms of seismic design philosophies, seismic isolation, and pushover analysis techniques, are followed by a brief treatment on the computer modeling of bridges.

Chapter 5 is a description of earthquake loading as applied to structures, whether buildings, bridges, or other. Baseline adjustment and filtering of ground motion, as well as the computation of various ground motion parameters and response spectra, are included. Requirements found in various ASCE design specifications are presented, along with a discussion on the importance of ground motion directionality and statistical considerations. Ground motion databases available to the engineer and scientist are identified followed by material on ground motion models, ground motion selection, and ground motion modification for structural analysis.

Chapter 6 provides the reader with example problems to solve for a clearer understanding of the design concepts presented in the book.

The appendices to the book include hand calculations corresponding to several examples presented. These hand calculations are better presented as appendices to retain the flow of the material presented in the main body.

2 Analysis Techniques for the Structural Engineer

Static structural analysis by any of the following methods may be accomplished. The method appropriate for a particular problem depends on the expected level of structural response. Nonlinear effects may be broadly categorized as (a) geometric and (b) material. Geometric nonlinearity refers to $P\Delta$ and $P\delta$ effects, while material nonlinearity refers to strains beyond the yield point. $P\Delta$ effects refer to moment and deflection amplification which occur when compression member ends experience relative translation. $P\delta$ effects refer to moment and deflection amplification which occur when no member end relative translation exists, but transverse loads between member ends exist in compression members.

- **First-Order Elastic Analysis**: neglects all nonlinearities and is typically a good indicator of service load conditions.
- **Second-Order Elastic Analysis**: $P\Delta$ and $P\delta$ effects are included, but no material nonlinearity is incorporated.
- **First-Order Inelastic Analysis:** Equilibrium is established on the undeformed structure and material nonlinearity is included.
- **Second-Order Inelastic Analysis**: Equilibrium is established on the deformed structure and material nonlinearity is included.

In addition to static methods, dynamic analysis methods are often required. These include response spectrum analysis, linear response history analysis, and nonlinear response history analysis.

The following sections provide further discussion and examples for these types of structural analysis.

2.1 FIRST-ORDER ELASTIC STRUCTURAL ANALYSIS

A first-order structural analysis is the most basic form of structural analysis studied in undergraduate structural mechanics courses. Equilibrium is established on the undeformed structure in a first-order analysis. No straining beyond the yield point is accounted for in elastic analysis.

Example 2.1-1: First-Order Elastic Analysis

The W40X277 beam shown is made from a new Grade 100 steel (F_y = 100 ksi, F_u = 110 ksi) and braced only at the ends and at the loads as indicated by the symbols. The beam is pinned at the left end with a roller at the right end.

The braces shown are effective for both torsional braces in resisting flexure and for weak axis compression buckling. The braces shown are not effective for strong axis buckling in compression.

Ignore the beam self-weight effects for hand calculations. Include the self-weight as part of the Dead Load in an IES VisualAnalysis (IES, 2020) computer model. Check the section against AISC 360-16 (American Institute of Steel Construction, 2016) Chapter H requirements using a first-order analysis.

The load combination to be addressed is taken from ASCE 7-16 for load and resistance factor design (LRFD) principles. The controlling load combination for this problem is $U = 1.2D + 1.6L$ (Figure 2.1.1).

$$R_u = 1.2 \times 125 + 1.6 \times 105 = 318 \text{ kips}$$

$$P_u = 1.6 \times 1,750 = 2,800 \text{ kips}$$

$$w_u = 1.2 \times 0.277 \text{ klf} = 0.332 \text{ klf}$$

The properties for the W40X277 are summarized next.

A = 81.5 in²	I_x = 21,900 in⁴	r_{TS} = 4.25 inches	d = 39.7 inches
S_x = 1,100 in⁴	h_o = 38.1 inches	t_w = 0.83 inches	r_x = 16.4 inches
J = 51.5 in⁴	C_w = 379,000 in⁶	b_f = 15.8 inches	t_f = 1.58 inches
Z_x = 1,250 in³	I_y = 1,040 in⁴	r_y = 3.58 inches	S_y = 132 in³
Z_y = 204 in³	b/t = 5.03	h/t_w = 41.2	

The section classification for compression is determined in accordance with Table B4.1a from AISC 360-16.

$$\lambda_f = \frac{b}{t} = 5.03 < 0.56\sqrt{\frac{29,000}{100}} = 9.54 \rightarrow \text{flange is nonslender}$$

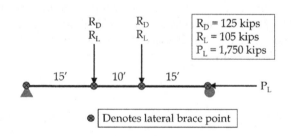

R_D = 125 kips
R_L = 105 kips
P_L = 1,750 kips

⊗ | Denotes lateral brace point

FIGURE 2.1.1   First-order elastic analysis example.

$$\lambda_w = \frac{h}{t_w} = 41.2 > 1.49\sqrt{\frac{29,000}{100}} = 25.4 \rightarrow \text{web is slender}$$

The section classification for flexure is determined in accordance with Table B4.1b from AISC 360-16.

$$\lambda_f = 5.03 < \lambda_{pf} = 0.38\sqrt{\frac{29,000}{100}} = 6.47 \rightarrow \text{flange is compact}$$

$$\lambda_w = 41.2 < \lambda_{pf} = 3.76\sqrt{\frac{29,000}{100}} = 64.03 \rightarrow \text{web is compact}$$

For the design compressive resistance, ϕP_n, the controlling effective length is required. The major-axis (x-axis) unbraced length for compression is 40 ft, with an effective length factor $K_x = 1.0$. The minor axis (y-axis) unbraced length is 15 ft for the end segments, with an effective length factor $K_y = 1.0$. The minor axis (y-axis) unbraced length is 10 ft for the middle segment, with an effective length factor $K_y = 1.0$. The applicable sections of AISC 360-16 are E3 and E7 (since the web is slender) for calculation of the design compressive resistance.

$$\left(\frac{KL}{r}\right)_x = \frac{1.0 \times 40 \times 12}{16.4} = 29.27$$

$$\left(\frac{KL}{r}\right)_{y-end} = \frac{1.0 \times 15 \times 12}{3.58} = 50.3 \leftarrow \text{controls end segments}$$

$$\left(\frac{KL}{r}\right)_{y-mid} = \frac{1.0 \times 10 \times 12}{3.58} = 33.5 \leftarrow \text{controls middle segment}$$

For compressive resistance of the end segments:

$$\frac{L_c}{r} = 50.3$$

$$F_e = \frac{\pi^2 \times 29,000}{(50.3)^2} = 113.2\,\text{ksi}$$

$$\frac{F_y}{F_e} = \frac{100}{113.2} = 0.883 < 2.25$$

$$F_{cr} = 0.658^{0.883} \times 100 = 69.1\,\text{ksi}$$

The web is a stiffened element, so for the provisions in AISC 360-16, Section E7, $c_1 = 0.18$ and $c_2 = 1.31$. Section E7 provides a means of determining the effective area, A_e, for sections with slender elements.

$$\lambda_w = \frac{h}{t_w} = 41.2 > \lambda_r \sqrt{\frac{F_y}{F_{cr}}} = 25.4\sqrt{\frac{100}{69.1}} = 30.56 \rightarrow A_e < A_g$$

$$F_{el} = \left(1.31 \times \frac{25.4}{41.2}\right)^2 \times 100 = 65.2 \, \text{ksi}$$

$$\frac{h_e}{h} = \left[1 - 0.18\sqrt{\frac{65.2}{69.1}}\right] \times \sqrt{\frac{65.2}{69.1}} = 0.802$$

$$h = \frac{h}{t_w} \times t_w = 41.2 \times 0.83 = 34.2 \, \text{in}$$

Determine the ineffective web area and subtract this from the gross area to determine the effective area for compressive resistance. The design compressive resistance can then be calculated.

$$A_e = 81.5 - (1 - 0.802) \times 34.2 \times 0.83 = 75.9 \, \text{in}^2$$

$$\phi P_n = 0.90 \times 75.9 \times 69.1 = 4{,}719 \, \text{kips, end segments}$$

Similar calculations for the middle segment give the design resistance for the middle segment.

$$F_e = \frac{\pi^2 \times 29{,}000}{(33.5)^2} = 254.7 \, \text{ksi}$$

$$\frac{F_y}{F_e} = \frac{100}{254.7} = 0.393 < 2.25$$

$$F_{cr} = 0.658^{0.393} \times 100 = 84.8 \, \text{ksi}$$

$$\lambda_w = \frac{h}{t_w} = 41.2 > \lambda_r \sqrt{\frac{F_y}{F_{cr}}} = 25.4\sqrt{\frac{100}{84.8}} = 27.6 \rightarrow A_e < A_g$$

$$F_{el} = \left(1.31 \times \frac{25.4}{41.2}\right)^2 \times 100 = 65.2 \, \text{ksi}$$

$$\frac{h_e}{h} = \left[1-0.18\sqrt{\frac{65.2}{84.8}}\right] \times \sqrt{\frac{65.2}{84.8}} = 0.738$$

$$A_e = 81.5 - (1-0.738) \times 34.2 \times 0.83 = 74.1\,\text{in}^2$$

$$\phi P_n = 0.90 \times 74.1 \times 84.8 = 5,653\,\text{kips, middle segment}$$

For the design flexural resistance, ϕM_n, it is necessary to consider the end segments and the middle segment of the beam since the unbraced length, as well as the lateral-torsional-buckling (LTB) modification factor, C_b, is different for each. This is illustrated in Figure 2.1.2. For determination of the moment diagram and the LTB modification factors, consider the concentrated loads only since the self-weight adds a relatively small amount to the flexural demand. This is an approximation in hand calculations, but most software will automatically account for self-weight.

$$M_u = 318\,\text{kips} \times 15\,\text{ft} = 4,770\,\text{ft} \cdot \text{kips}$$

The unbraced length limits, L_p and L_r, defining the transition zones for flexural resistance can be determined using the provisions in Chapter F, Section F2 of AISC 360-16, since the cross section is compact for flexure.

$$L_p = 1.76 r_y \sqrt{\frac{E}{F_y}} = 1.76(3.58)\sqrt{\frac{29,000}{100}} = 107.3\,\text{in} = 8.94\,\text{ft}$$

$$\frac{Jc}{S_x h_o} = \frac{51.5 \times 1.0}{1,100 \times 38.1} = \frac{1}{813.8}$$

$$6.76\left(\frac{0.7F_y}{E}\right)^2 = 6.76\left(\frac{0.7 \times 100}{29,000}\right)^2 = \frac{1}{25,389}$$

$$L_r = 1.95\, r_{TS} \frac{E}{0.7F_y} \sqrt{\frac{Jc}{S_x h_o} + \sqrt{\left(\frac{Jc}{S_x h_o}\right)^2 + 6.76\left(\frac{0.7F_y}{E}\right)^2}}$$

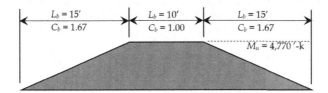

| $L_b = 15'$ | $L_b = 10'$ | $L_b = 15'$ |
| $C_b = 1.67$ | $C_b = 1.00$ | $C_b = 1.67$ |

$M_u = 4,770''\text{-k}$

FIGURE 2.1.2 Braced segments (first-order analysis example).

$$L_r = 1.95(4.25)\frac{29,000}{0.7\times100}\sqrt{\frac{1}{813.8}+\sqrt{\left(\frac{1}{813.8}\right)^2+\frac{1}{25,839}}} = 300 \text{ in} = 25 \text{ ft}$$

With the actual unbraced length of each segment between L_p and L_r, the design flexural resistance is the resistance factor for flexure, $\phi = 0.90$, multiplied by the nominal resistance given by Equation F2-2 of AISC 360-16.

$$\phi M_n = 0.90C_b\left[M_p - (M_p - 0.7F_yS_x)\left(\frac{L_b-L_p}{L_r-L_p}\right)\right]\leq 0.90M_p$$

$$M_p = 100\times1,250 = 125,000 \text{ in}\cdot\text{k} = 10,417 \text{ ft}\cdot\text{kips}$$

$$0.7F_yS_x = 0.7\times100\times1,100 = 77,000 \text{ in}\cdot\text{k} = 6,417 \text{ ft}\cdot\text{kips}$$

For the end segments, with $L_b = 15$ ft and $C_b = 1.67$:

$$\phi M_n = 0.90(1.67)\left[10,417-(10,417-6,417)\left(\frac{15-8.94}{25-8.94}\right)\right]$$

$$= 13,388 > 0.90M_p = 9,375$$

$$\phi M_n = 0.90M_p = 9,375 \text{ ft}\cdot\text{k}$$

For the middle segment, with $L_b = 10$ ft and $C_b = 1.00$:

$$\phi M_n = 0.90(1.00)\left[10,417-(10,417-6,417)\left(\frac{10-8.94}{25-8.94}\right)\right]$$

$$= 9,138 < 0.90M_p = 9,375$$

$$\phi M_n = 0.90M_p = 9,138 \text{ ft}\cdot\text{k}$$

For combined axial compression and major-axis flexure, AISC 360-16 Chapter H requirements apply. While each of the segments, end and interior, are subjected to the same axial force and the same maximum moment, the design resistances are different for the segments. So each segment will be checked.

For the end segments:

$$\frac{P_u}{\phi P_n} = \frac{2,800}{4,719} = 0.593 > 0.20$$

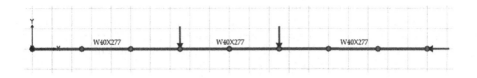

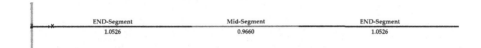

FIGURE 2.1.3   IES VisualAnalysis results (first-order analysis example).

$$0.593 + \frac{8}{9}\left(\frac{4,770}{9,375}\right) = 1.045 > 1.00; \text{ no good-end segments}$$

For the middle segment:

$$\frac{P_u}{\phi P_n} = \frac{2,800}{5,653} = 0.495 > 0.20$$

$$0.495 + \frac{8}{9}\left(\frac{4,770}{9,138}\right) = 0.959 < 1.00; \text{ OK middle segment}$$

The IES VisualAnalysis model results are summarized in Figure 2.1.3, with self-weight neglected and included in the second and third plots, respectively. These confirm the hand calculations for a first-order analysis. The conclusion from a first-order analysis is that the end segments are about 5% overloaded, and the middle segment meets the design requirements for axial force-flexure interaction.

2.2   SECOND-ORDER ELASTIC STRUCTURAL ANALYSIS

A second-order elastic analysis incorporates geometric nonlinearity but not material nonlinearity. Equilibrium is established on the deformed structure.

Example 2.2-1: Second-Order Elastic Analysis

The problem presented in Section 2.1 is now solved using a second-order elastic analysis by the Direct Analysis method of AISC 360-16, Chapter C, along with the approximate second-order analysis method in Appendix 8. The Direct Analysis method requires consideration of (a) all sources of deformation, (b) stiffness reduction to account for residual stresses, and (c) notional loads (or explicit imperfection modeling) to account for out-of-straightness and out-of-plumb imperfections. The method is discussed in detail in Section 3.2.

The first step in performing the direct analysis solution is to establish the level of stiffness reduction required for each segment. Refer to Section C2 of AISC 360-16.

$$P_{ns} = F_y A_e = 100 \times 75.9 = 7{,}590 \text{ kips, end segments}$$

$$\frac{P_u}{P_{ns}} = \frac{2{,}800}{7{,}590} = 0.369 < 0.500 \rightarrow \tau_b = 1.00, \text{ end segments}$$

$$P_{ns} = F_y A_e = 100 \times 74.1 = 7{,}410 \text{ kips, middle segment}$$

$$\frac{P_u}{P_{ns}} = \frac{2{,}800}{7{,}410} = 0.378 < 0.500 \rightarrow \tau_b = 1.00, \text{ middle segment}$$

$$\rightarrow EI^* = 0.80EI$$

Appendix 8 of AISC 360-16 provides a means of performing the approximate second-order analysis permitted by Chapter C in the Direct Analysis Method. With no relative translation of the member ends, the problem is a $P\delta$ problem, and the amplifier, B_1, is needed. With transverse loads between the supports, $C_m = 1.0$.

$$P_{e1} = \frac{\pi^2 (0.80 \times 29{,}000 \times 21{,}900)}{(40 \times 12)^2} = 21{,}764 \text{ kips}$$

$$B_1 = \frac{1.0}{1.0 - \dfrac{2{,}800}{21{,}746}} = 1.148$$

From the Commentary to Section C2.2 of AISC 360-16, initial imperfections may be modeled assuming an out-of-straightness equal to L/1,000 or by the application of notional loads. For this example, consider an out-of-straightness equal to (40 × 12)/1,000 = 0.48 inches at midspan. Then the first-order moment is determined by adding this effect to the previously calculated factored moment due to the applied vertical loads of 4,770 ft·k.

$$M_{nt} = 4{,}770 + 2{,}800 \times \frac{0.48}{12} = 4{,}882 \text{ ft·k}$$

The second-order moment is simply the amplification factor, B_1, multiplied by the first-order moment. The axial load-flexure interaction equation from Chapter H of AISC 360-16 may then be applied to assess the adequacy of the middle segment. There is no amplification applied to axial load when $P\delta$ effects are the only second-order effects (no $P\Delta$ effects, $B_2 = 0$). Since $K = 1$ has already been used to determine axial resistance, no adjustment in ϕP_n is required for this Direct Analysis solution.

$$M_u = 1.148 \times 4{,}882 = 5{,}603 \text{ ft} \cdot \text{k}$$

For the middle segment:

$$\frac{P_u}{\phi P_n} = \frac{2{,}800}{5{,}653} = 0.495 > 0.20$$

$$0.495 + \frac{8}{9}\left(\frac{5{,}603}{9{,}138}\right) = 1.040 > 1.00; \text{ no good-middle segment}$$

The IES VisualAnalysis results are shown in Figure 2.2.1. These confirm the hand calculations. While a first-order analysis indicates 5% overload in the end segments and no overload in the middle segment, the more accurate Direct Analysis

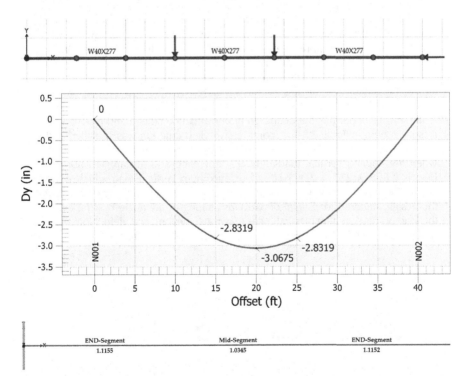

FIGURE 2.2.1 IES VisualAnalysis results (second-order elastic analysis example). (a) Model with loads (b) Deflected shape. (c) Design ratios.

method solution indicates 12% overload in the end segments and 4% overload in the middle segment. Figure 2.2.1 also includes second-order deflections, which are about 43% larger than deflections from a first-order analysis under factored loads. This should make it clear that deflection amplification and moment amplification are far from equal in many cases.

2.3   SECOND-ORDER INELASTIC STRUCTURAL ANALYSIS

A second-order inelastic analysis incorporates both geometric and material nonlinearities. Equilibrium is established on the deformed structure. Various nonlinear material stress-strain relationships are available in software. Some examples are (a) the Mander model for confined concrete, (b) the Ramberg-Osgood model for steel, (c) the Menegotto-Pinto steel model, and (d) simple, bilinear models. Simple bilinear models are often used to model seismic isolation devices as link-type elements.

Example 2.3-1: Second-Order Inelastic Analysis

Seismostruct (Seismosoft, 2020) has been used to solve the problem presented in Sections 2.1 (first-order elastic) and 2.2 (second-order elastic) using a second-order inelastic analysis (Figure 2.3.1).

For a second-order inelastic analysis, it is typically required to increment the loads until a failure condition has been reached. For the example considered here, note that R_u (vertical factored load at each of two points on the beam) is 318 kips and P_u (horizontal, compressive load applied at the end) is 2,800 kips. Vertical loads equal to 1.0 kips and a horizontal load equal to 8.805 kips (2,800/318) are applied as initial loads in the Seismostruct model, and a target load factor equal to 318 is specified. A reduced stiffness as required by AISC 360 is incorporated by specifying $E = 0.997 \times 0.80 \times 29,000 = 23,130$ ksi.

Seismostruct permits the definition of performance criteria and reports results as shown here in Table 2.3.1. A compressive yield criterion was specified with a strain equal to $100/29,000 = 0.003448$. First yield occurs in the example problem when the applied vertical loads are 272 kips each and the applied horizontal load is $8.805 \times 272 = 2,395$ kips. The maximum vertical load which may be applied before failure by instability is 368 kips, indicating acceptable response at the design, factored load level of 318 kips as long as adequate bracing exists at the load points such that plastic hinges may rotate beyond first yield. Note that the design resistance is $0.90 \times 368 = 331$ kips, which is greater than 318 kips.

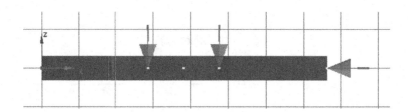

FIGURE 2.3.1   Seismostruct model for inelastic second-order analysis.

TABLE 2.3.1
Beam Example Performance Criteria

Load Factor	Criterion	Beam Segment	Location
272	Compression yield	Middle segment	Midspan
273	Compression yield	Left segment	Under load
273	Compression yield	Right segment	Under load
368	Overall instability	Middle segment	Midspan

Example 2.3-2: Second-Order Inelastic Analysis of a Frame

Structural analysis by second-order principles with inelasticity incorporated can be useful for frames as well as beams. Seismostruct is again used to model the frame indicated in Figure 2.3.2. The goal for this example is to determine the maximum load, P_u, which may be applied to the frame. Hand calculations are supplemented with a description of a Seismostruct model and results. From hand calculations, beam, sway, and combined mechanisms are investigated.

The plastic moment for the beam is M_p = 50 ksi × 373 in³/12 = 1,554 ft·kips. From hand calculations (Appendix A), the combined mechanism controls and P_u = 16/75 × 1,554 = 332 kips.

One analysis type available in Seismostruct is a static pushover analysis, which was applied here and is useful for second-order inelastic analysis. Applied loads of 1 kip were applied vertically and horizontally with a target load factor of 400 (larger than the estimated capacity of 332 kips). Performance criteria results are shown in Table 2.3.2 and interpreted as follows:

First yield occurs at the bottom of the right column at a load of 251 kips. The next yield occurs at both the right end of the right beam segment and the top of the right column at a load of 284 kips. At a load equal to 311 kips, yielding occurs

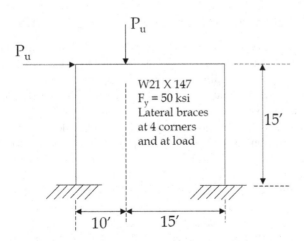

FIGURE 2.3.2 Second-order inelastic analysis of a frame.

TABLE 2.3.2

Frame Example Performance Criteria

Load Factor	Criterion	Element	Location
251	Compression yield	Right column	Base
284	Compression yield	Right beam segment	Right end
284	Compression yield	Right column	Top
311	Compression yield	Left beam segment	Right end
311	Compression yield	Right beam segment	Left end
313	Compression yield	Left column	Base
345	Fracture	Left column	Base

at the vertical load (the right end of the left beam segment and the left end of the right beam segment). Finally, yielding occurs at the base of the left column when the applied loads are each 313 kips. This analysis confirms the combined mechanism found through hand calculations. First yield of the left column base at a load of 313 kips roughly corresponds to the load estimated from hand calculations of 332 kips. Establishing a performance criteria of, say, strain equal to 20 times the yield strain = 0.0345 could be specified to attempt an estimate of the load at which the final hinge forms at the base of the left column. Such a criterion was added and the indicated collapse load was found to be 345 kips; near the hand-calculated value of 332 kips.

The progressive formation and plastic rotation of plastic hinges requires compact sections and sufficient bracing. If the given frame is adequately braced at the corners and at the load points, AISC 360-16 provisions are useful in determining whether or not such an analysis would be permitted for structural design. Hand calculations demonstrating satisfaction of requirements in accordance with AISC provisions are presented in Appendix A.

2.4 LINEAR ELASTIC RESPONSE SPECTRUM ANALYSIS

One of the first topics covered in structural dynamics for civil engineers is the concept of modal response spectrum analysis. This is still the "default" method for many design offices for typical building structures. In response spectrum analysis, a structure with "n" degrees of freedom is solved as the linear combination of "n" single-degree-of-freedom systems through application of modal orthogonality properties.

Modern codes and specifications are typically very detailed in defining the design response spectrum. One generic response spectrum shape, which has been used in many specifications, is shown in Figure 2.4.1. The spectrum, spectral acceleration versus structure period, is entirely defined by the parameters S_{DS}, S_{D1}, T_S, T_O, and T_L. Chapter 5 contains an extensive coverage of response spectrum definitions in accordance with various codes and specifications. Design response spectra are typically defined at a damping level equal to 5% of the critical.

The dynamic properties of a structure include the natural frequencies (equal in number to the number of degrees of freedom in the structure), the mode shapes,

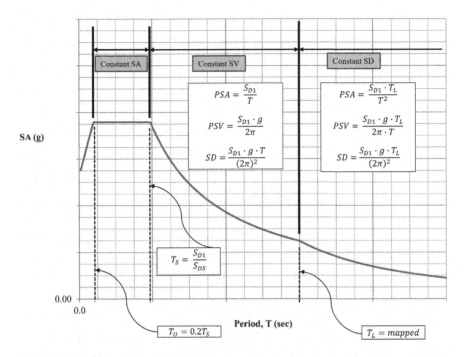

FIGURE 2.4.1   Generic design response spectrum.

effective modal masses, and modal participation factors. It is typically not necessary to include all frequencies to achieve an accurate solution.

Numerical software, such as Scilab (ESI Group, 2020), may be used to extract the natural frequencies (ω) and mode shapes (ϕ) for a dynamic system defined by mass matrix, M, and stiffness matrix, K. The squares of the natural frequencies are given by the eigenvalues of $M^{-1}K$, and the mode shapes are given by the eigenvectors of $M^{-1}K$. For a complete and comprehensive treatment of structural dynamics, consult one of the excellent textbooks on the subject (Chopra, Earthquake Dynamics of Structures, 2005) (Clough & Penzien, 1975) (Chopra, Dynamics of Structures, 2016) (Paz, 2007). Equations 2.4-1 through 2.4-7 summarize some of the important dynamic properties and results from modal superposition analysis.

$$GM = \phi^T M \phi, \text{ generalized mass matrix} \qquad (2.4\text{-}1)$$

$$GK = \phi^T K \phi, \text{ generalized stiffness matrix} \qquad (2.4\text{-}2)$$

$$\omega_i^2 = \frac{GK(i,i)}{GM(i,i)} \qquad (2.4\text{-}3)$$

$$EMM = \frac{\left[\phi^T M r\right]^2}{\phi^T M \phi}, \text{ effective modal mass} \qquad (2.4\text{-}4)$$

$$\Gamma = \frac{\phi^T M r}{\phi^T M \phi}, \text{ modal participation factor} \qquad (2.4\text{-}5)$$

$$Y_i = \Gamma_i \left(SD_i\right)\phi_i, \text{ modal displacement} \qquad (2.4\text{-}6)$$

$$A_i = \Gamma_i \left(SA_i\right)\phi_i, \text{ modal acceleration} \qquad (2.4\text{-}7)$$

Spectral displacement, *SD*, is often determined from the spectral acceleration, *SA*, as $SD = SA/\omega^2$. In these equations, "*r*" is a column vector consisting of 1's for each degree of freedom in the direction of applied loading (typically horizontal degrees of freedom) and zeroes for all other degrees of freedom.

Once response values (displacement and acceleration) for each mode have been determined, they must be combined in some manner. Various methods for these combination procedures exist. With an "ABSUM" combination rule, one simply takes the absolute value of response at each mode and adds each contributing modal response. In the "SRSS" combination rule, one takes the square root of the sum of the squares for each degree of freedom and for each mode, adds and takes the square root. The "CQC" combination (complete quadratic combination) rule accounts for cross-correlations in each modal response and is often required when closely spaced modes exist.

Example 2.4-1: Modal Response Spectrum Analysis

The shear building structure shown in Figure 2.4.2 is to be located at a site with the design response spectrum key points (refer to Chapter 5 for a detailed discussion of design response spectra):

- $S_{DS} = 0.750$ g
- $S_{D1} = 0.600$ g
- $T_L = 4$ sec

Find:

1. Natural frequencies and mode shape plots
2. Percent of mass participating in each mode
3. ABSUM and SRSS accelerations at each floor
4. ABSUM and SRSS displacements at each floor
5. ABSUM and SRSS drift for each story

Scilab (ESI Group, 2020) has been used to perform all necessary eigen-analysis steps. The commands and ensuing results are displayed in Figure 2.4.3.

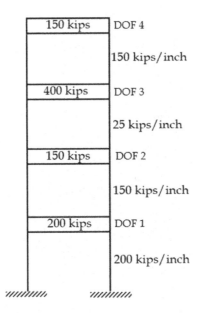

FIGURE 2.4.2   Shear building example for response spectrum analysis.

Once the eigenvalues and eigenvectors have been obtained, Excel is convenient for performing the remaining operations and plotting the mode shapes. Figure 2.4.4 shows the orthogonality calculations in Excel. Figure 2.4.5 summarizes the modal analysis results from Excel. Figures 2.4.6 through 2.4.9 are mode shape plots generated in Excel.

Several observations may be made from the results.

1. For dof's 1 and 2, displacements are higher for mode 1 than for mode 2.
2. For dof's 1 and 2, accelerations are higher for mode 2 than for mode 1.
3. The "soft" third story results in lower accelerations for dof's 3 and 4, but very large displacements for dof's 3 and 4, and a very large drift at the third dof.
4. The sum of EMMs is equal to the sum of actual masses.
5. ABSUM is typically overly conservative due to the inherent assumption in ABSUM that all modal peaks occur simultaneously. If each frequency is an integer multiple of the preceding one, and if there is no damping, then ABSUM may be appropriate.
6. SRSS is typically a "reasonable" approximation of the true solution.
7. The ABSUM or SRSS floor force may be taken equal to the floor mass multiplied by the ABSUM or SRSS acceleration.
8. The ABSUM or SRSS story shear may be taken equal to the story stiffness multiplied by the ABSUM or SRSS drift.

```
-->K=[350 -150 0 0; -150 175 -25 0; 0 -25 175 -150; 0 0 -150 150]
 K  =

   350.   - 150.    0.       0.
 - 150.     175.   - 25.      0.
    0.     - 25.    175.    - 150.
    0.       0.    - 150.    150.

-->M=[0.518 0 0 0; 0 0.3885 0 0; 0 0 1.0360 0; 0 0 0 0.3885]
 M  =

   0.518    0.        0.        0.
   0.       0.3885    0.        0.
   0.       0.        1.036     0.
   0.       0.        0.        0.3885

-->MI=M^(-1)
 MI  =

   1.9305019    0.           0.           0.
   0.           2.5740026    0.           0.
   0.           0.           0.9652510    0.
   0.           0.           0.           2.5740026


-->A=MI*K

 A  =

   675.67568   - 289.57529     0.           0.
 - 386.10039     450.45045   - 64.350064    0.
    0.          - 24.131274    168.91892  - 144.78764
    0.            0.         - 386.10039    386.10039

-->[Vect,Val]=spec(A)
 Val  =

   916.71451      0            0             0
   0            213.02075      0             0
   0              0           13.215561      0
   0              0            0           538.19461

 Vect  =

 - 0.7682395   - 0.5256918    0.0716148     0.0550340
   0.6394729   - 0.8398988    0.1638328     0.0261284
 - 0.0240198     0.0552122    0.6834945   - 0.3658314
   0.0174780     0.1231657    0.7077185     0.9286851
```

FIGURE 2.4.3   Scilab session for eigenvalue analysis.

$\phi^T_m\phi$	0.6917	0.0000	0.0000	0.0000
	0.0000	0.4263	0.0000	0.0000
	0.0000	0.0000	0.4755	0.0000
	0.0000	0.0000	0.0000	0.4653
$\phi^T_k\phi$	9.1406	0.0000	0.0000	0.0000
	0.0000	90.8025	0.0000	0.0000
	0.0000	0.0000	255.9378	0.0000
	0.0000	0.0000	0.0000	426.5504

FIGURE 2.4.4 Excel orthogonality calculations.

	ABSUM	SRSS
Acc, g	0.7048	0.5030
	0.9921	0.7548
	0.4380	0.3751
	0.5266	0.4007
	ABSUM	SRSS
Y, in	2.0542	1.4098
	3.9990	2.9210
	10.9604	10.8627
	11.4646	11.2490
	ABSUM	SRSS
Drift, in.	2.0542	1.4098
	2.1210	1.5550
	9.7521	8.3783
	0.5266	0.4007

FIGURE 2.4.5 Modal analysis results in Excel.

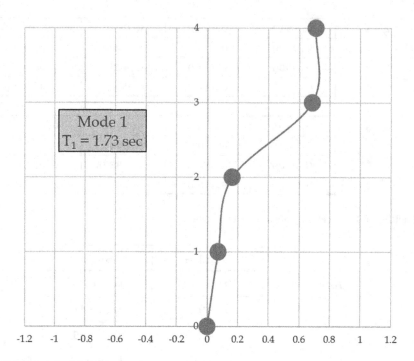

FIGURE 2.4.6 Mode shape 1 for example shear building.

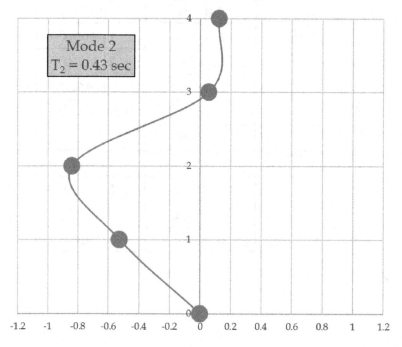

FIGURE 2.4.7 Mode shape 2 for shear building example.

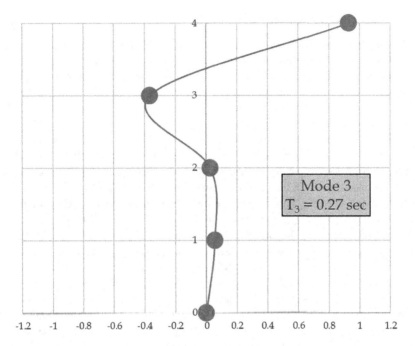

FIGURE 2.4.8   Mode shape 3 for example shear building.

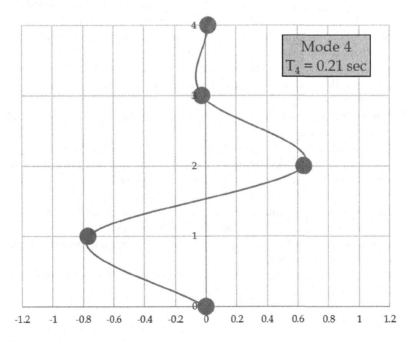

FIGURE 2.4.9   Mode shape 4 for example shear building.

2.5   RESPONSE HISTORY ANALYSIS

When response spectrum analysis is not permitted by the governing design criteria or may not be appropriate, it becomes necessary to employ response history
techniques, which may be either linear or nonlinear in nature. In order to perform
response history analyses for seismic effects, it is first necessary to select and modify
appropriately a suite of ground motions to be used as applied loadings in the analysis
and design.

The design, or "target," response spectrum to be used for ground motion modification may or may not be the same as the code-based response spectrum. Code-based
response spectra are typically uniform hazard response spectra (UHRS) or risk-targeted response spectra (RTRS). UHRS usually are envelopes of spectra from ground
motion models (GMM) evaluated over a range of potential earthquake moment magnitudes and distances. Hazard analysis involves identification of the faults that contribute to the hazard at the site of interest. It has been generally accepted that the
UHRS is a conservative basis for a target response spectrum to be used in ground
motion selection and modification (NEHRP Consultants Joint Venture, 2011). An
alternative is the conditional mean spectrum (CMS), in which pseudo-spectral-accel-
eration (PSA) ordinates at periods other than a specified period (or range of periods)
are not taken as UHRS values, but as values estimated based on the condition of a
UHRS value at the period of interest. The CMS matches the UHRS at the specified
period(s), and is below the UHRS at all other periods. The CMS has been suggested
as a more realistic target for ground motion modification.

AASHTO design response spectra for bridges are not the same as ASCE 7-16
design response spectra for buildings. AASHTO design spectra have historically
been based on the uniform hazard, geometric mean of two horizontal components.
ASCE 7-16 spectra are risk-targeted and based on the maximum direction of two
horizontal components rotated through all non-redundant angles. As of the writing
of this section, Research is ongoing that will eventually alter the nature of design
ground motions for bridges.

Regardless of the target chosen, it is essential that the engineer be aware of the
target basis concerning directionality and hazard. There are many different definitions used to define a single spectral value at each period given that the ground
motion spectra are dependent upon instrument orientation and three components
are recorded. Some of the more common definitions (there are many others) include:

GMAR (as-recorded geometric mean) – the geometric mean PSA of the two
   horizontal as-recorded ground motions at each period
RotD50 – period-dependent median component PSA rotated through all non-
   redundant rotation angles
RotD100 – period-dependent maximum component PSA rotated through all
   non-redundant rotation angles

Detailed definitions of these and other intensity measures may be found in the literature (Boore D. M., August 2010), (Boore, Watson-Lamprey, & Abrahamson, August 2006), (Stewart, et al., 2011).

Chapter 5 includes an extensive coverage of ground motion selection and modification procedures for structural analysis, as well as detailed guidance on the development of design response spectra based on multiple specifications.

Response history analysis of building structures may be either linear or nonlinear in nature. Once a target response spectrum has been selected, a set of ground motion records is developed. This includes selection criteria as well as modification criteria. Sources for ground motion records include:

- PEER NGA-West2 Database (Pacific Earthquake Engineering Research Center, 2020)
- Center for Engineering Strong Motion Data (USGS and California Geological Survey, 2011)
- COSMOS Virtual Data Center (Consortium of Organizations for Strong Motion Observation Systems, 2007)
- Engineering Strong Motion Database (Luzi, Puglia, & E., ORFEUS WG5 (2016). Engineering Strong Motion Database, 2016)

Initial selection criteria for candidate records typically include specified ranges for each of the following:

1. Moment magnitude, M_W
2. Shear wave velocity in the upper 30 m, V_{S30}
3. Distance from fault, R
4. Tectonic regime (active tectonic, stable continent, subduction)
5. Presence (or absence) of velocity pulse in the ground motion
6. Match to spectral shape

There are several metrics used for fault distance: R_{jb} (Joyner-Boore distance), R_{epi} (epicentral distance), R_{hyp} (hypocentral distance), etc.

Modification procedures include:

- Amplitude scaling, either at a single period of interest or, more likely, to minimize mean-square-error between the suite mean and the target over a range of periods of interest
- Spectral matching by the addition of wavelets in the time domain
- Spectral matching by Fourier amplitude manipulation in the frequency domain

Of these, amplitude scaling is typically preferred. The online PEER NGA-West2 Ground Motion Database (Pacific Earthquake Engineering Research Center, 2020) uses this method. SeismoMatch (SeismoSoft, 2020) provides a means of applying time domain spectral matching. SeismoArtif (SeismoSoft, 2020) employs frequency-domain spectral matching. Chapter 5 contains detailed discussion of the ground motion selection and modification procedures. Sufficient for now is a realization that a hazard deaggregation is one of the first steps in performing ground motion selection for structural analysis and design.

Ground motion selection often requires deaggregation of the seismic hazard at a site to determine the contributing (M_w, R) pairs to the hazard. This may be accomplished using the USGS data applications (US Geological Survey, 2020).

Example 2.5-1: Seismic Hazard Deaggregation

A hypothetical building project near Memphis, Tennessee, is the basis for this example. The USGS Unified Hazard Tool (US Geological Survey, 2020) was used to locate the site at latitude 35.172°, longitude −90.071°. Figure 2.5.1 is a screenshot taken from the online tool.

The UHRS values corresponding to a mean recurrence interval (MRI) of 2,500 years were obtained for the site. These are summarized in Table 2.5.1. While AASHTO criteria are *GeoMean*-based for a 1,000-year MRI, ASCE criteria for buildings are *RotD100*-based for a 2,500-year MRI. Further, ASCE target spectra are risk-targeted (RTRS), while AASHTO target spectra are UHRS based.

The data sets available at the USGS Unified Hazard Tool are numerous. For this example, the "*Dynamic: Conterminous US 2014 (Update) (V4.2.0)*" has been adopted. Deaggregation results in terms of modal M, R (magnitude, distance) at control periods of PGA (peak ground acceleration), S_S (pseudo-spectral acceleration at a period of 0.20 sec), and S_1 (pseudo-spectral acceleration at a period of 1.0 sec) are shown in Table 2.5.1. The deaggregation was performed for the B/C site class boundary. The site is unique in that the modal M, R pairs are, for all practical purposes, identical at all three control periods. For actual building design and target spectra development, the PGA, S_S, and S_1 values would need to be modified for three effects.

(a) Conversion from UHRS to RTRS basis by application of appropriate risk factors, which depend on the region in which the project is located
(b) Conversion to *RotD100* from geometric mean by the application of appropriate amplification factors
(c) Convolution of B/C boundary accelerations to the surface using site class factors or site response analysis

For the selection of candidate records, it is typically necessary only to identify the M, R pairs contributing significantly to the seismic hazard at the project site. A large set of candidate records should first be selected based on deaggregation. Then, using procedures in Chapter 5, the most appropriate candidates can be identified and properly modified for use in structural analysis. It would be reasonable to identify 100 record pairs as candidates in order to develop a final set of 11–14 record pairs.

The next step in response history analysis would be to transfer the B/C boundary accelerations to the surface through the application of appropriate site factors or a

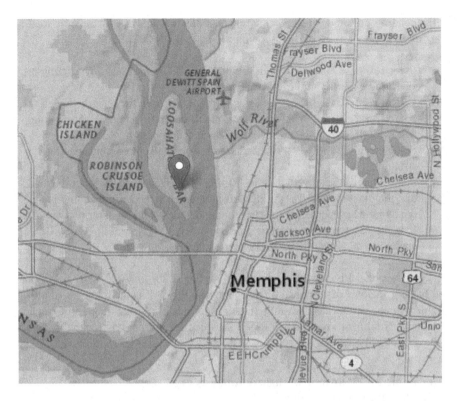

FIGURE 2.5.1   Site location for example 2.5-1 (USGS).

TABLE 2.5.1
Seismic Hazard Deaggregation

Parameter	Modal (*M, R*)	B/C Boundary PSA
2,500 year *PGA*	7.55, 51 km	0.635 g
2,500 year S_S	7.54, 52 km	1.102 g
2,500 year S_1	7.55, 52 km	0.316 g

---

site response analysis. See Chapter 5 for site factor alternatives and Section 2.6 for a discussion of site response analysis.

Ground motion candidates for use in response history analysis for the site would be selected based on the deaggregation given in Table 2.5.1. Candidate records would be modified by scaling or spectral matching, also covered in Chapter 5. One of the most complete and up-to-date references for ground motion selection and modification is freely available (NEHRP Consultants Joint Venture, 2011), and engineers performing this type of work will find it a valuable resource.

Depending on the desired results, response history analysis for seismic loading may include one or more of the following:

(a) Static response history analysis – typically used as a means of predicting or verifying experimental cyclic load testing
(b) Dynamic response history analysis – frequently used for seismic design of complex or critical structures; the typical loading is support acceleration histories; used in both seismic and blast-resistant design
(c) Static pushover analysis – often used to estimate displacement capacity of a structure
(d) Static adaptive pushover analysis – similar to pushover analysis, but the load distribution is modified as the analysis progresses in recognition of the changing dynamic properties as the structure undergoes inelastic deformations
(e) Incremental dynamics analysis – a set of analyses of increasing intensity is applied to the structure up to a specified target limit state

PRISM (National Research Foundation of Korea, 2020) may be used to introduce the basic principles of response history analysis. PRISM includes the capability to perform linear or nonlinear response history analysis of a single-degree-of-freedom oscillator. Nonlinear analysis in PRISM includes several options for the hysteretic behavior to be modeled.

Example 2.5-2: Response History Analysis of a Single-degree-of-Freedom Oscillator

Use PRISM to perform nonlinear response history analysis of a bilinear single-degree-of-freedom oscillator with the following properties. The initial elastic component of damping for this problem is taken equal to 1 percent of critical.

$W = 9,896$ kips $K_i = 2,395$ k/in $\alpha = 0.02$
$F_y = 2,806$ kips $S_{DS} = 0.917$ g $S_{D1} = 0.655$ g

Ground motions to be applied to the structure have been selected and modified based on the control points of the design response spectrum, S_{DS} and S_{D1}. The period range of interest has been set equal to 0.20–2.00 sec. Ground motions are shown in Table 2.5.2. Ground motions for this example were obtained from the NGA Subduction Database Preliminary Ground Motion Suite (Public Release) at the B. John Garrick Institute for the Risk Sciences:

risksciences.ucla.edu/nhr3/gmdata/preliminary-nga-subduction-records

Each of the two horizontal components for each record was applied separately and the response for that record taken equal to the geometric mean of the two horizontal responses.

Figure 2.5.2 shows the suite median spectrum with the design pseudo-acceleration spectrum. Table 2.5.3 summarizes the results of the nonlinear response history analysis from PRISM.

TABLE 2.5.2
Ground Motions for Response History Analysis

NGA Sub RSN	EQ	M_w	Year	SF	Station
4028438	Tokachi-oki	8.29	2003	5.767	AOM005
6001025	South.Peru	8.41	2001	1.862	POCON
4000244	Tohoku	9.12	2011	3.068	E40
4032540	Tokachi-oki	8.29	2003	5.193	51109
4001240	Tohoku	9.12	2011	4.459	YMT008
4000684	Tohoku	9.12	2011	4.565	YMTH14
4000348	Tohoku	9.12	2011	3.064	CHBH14
4032536	Tokachi-oki	8.29	2003	8.586	456
3001956	Michoacan, Mexico	7.99	1985	3.189	CALE
3001954	Michoacan, Mexico	7.99	1985	9.454	ATYC
6001378	Iquique	8.15	2014	11.254	PB04

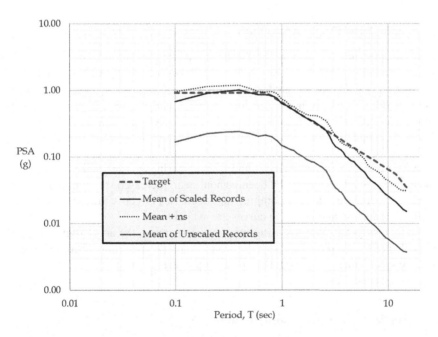

FIGURE 2.5.2 Suite median and target *PSA* for response history analysis.

The median displacement, the geometric mean displacement of the 11 record suite, is 10.40 cm. While this example is simple in nature, the process for a multi-degree-of-freedom system would be the same, with much more data through which to sift for results.

TABLE 2.5.3
PRISM Results for Nonlinear Response History Analysis

Record	D_{H1}, cm	D_{H2}, cm	D_{GM}, cm
4028438	16.94	12.55	14.58
6001025	9.10	11.33	10.15
4000244	12.96	13.04	13.00
4032540	16.20	12.57	14.27
4001240	12.06	7.89	9.75
4000684	7.98	11.87	9.73
4000348	12.05	11.20	11.62
4032536	11.56	10.40	10.96
3001956	14.02	7.12	9.99
3001954	15.33	7.70	10.86
6001378	1.68	10.86	4.27
Geo Mean	10.42	10.38	10.40

2.6   SITE RESPONSE ANALYSIS

For critical structures, and when subsurface conditions are outside of the range recommended for code-based site factor application to transfer bedrock motions to the surface, a site response analysis is required. Site response analysis may be linear, equivalent linear, or nonlinear. Site response analysis may be performed using appropriate ground motion accelerograms or may be based on random vibration theory. One open source tool that may be useful in site response is Strata (Kottke & Rathje, 2020).

As a soil mass is subjected to ground shaking during an earthquake, damping and shear modulus values eventually become nonlinear. Increased shear strain in the soil produces increased effective damping and decreases effective stiffness. Numerous modulus reduction and damping curves are available.

For site response analysis using input accelerograms, the accelerograms should be scaled or matched to align with the design response spectrum at bedrock, not the surface spectrum, since the intent is to develop a surface spectrum from the underlying bedrock spectrum.

Example 2.6-1: Site Response Using Strata

A 30 m thick uniform profile has a shear wave velocity V_S = 170 m/sec and density γ_{tot} = 18 kN/m³. Shear wave velocity at bedrock is taken equal to 3,000 m/sec for this example. The vertical, horizontal, and average confining pressures at mid-height of the layer (assuming a horizontal pressure coefficient equal to 0.50) are:

$$\sigma'_v = 18\,\text{kN/m}^3\,(15\,\text{m}) = 270\,\text{kPa}$$

$$\sigma'_h = 270\,\text{kPa}\,(0.50) = 135\,\text{kPa}$$

$$\sigma'_m = \frac{1}{3}(135 + 135 + 270) = 180\,\text{kPa} = 1.78\,\text{atm}$$

The design response spectrum at bedrock is taken as a geometric mean, UHRS near Memphis Tennessee having a mean recurrence interval equal to 2,500 years. The USGS Unified Hazard Tool (United States Geological Survey, 2020) was used to determine the following site class B/C boundary acceleration values (the US 2104 update (v4.2.0) data set was selected):

- Coordinates: 35.142° North, 90.051° West
- PGA: 0.601 g Modal M, R = 7.55, 55 km Mean M, R = 6.90, 38 km
- S_S = : 1.047 g Modal M, R = 7.55, 55 km Mean M, R = 7.16, 46 km
- S_1 = : 0.302 g Modal M, R = 7.55, 55 km Mean M, R = 7.50, 62 km

SigmaSpectra was used to develop a set of 11 ground motion pairs with amplitude scaling. A target logarithmic standard deviation equal to 0.6 across all periods of interest was adopted. This is based on ASCE 7-16 (American Society of Civil Engineers, 2017) Section 21.2.1.2. A broad period range of 0.2–6.0 sec was selected. A relatively wide magnitude range of M_W = 7.00–8.00 centered approximately on the modal M_W = 7.55 was used to develop the ground motion suite. Only records from site classes A, B, or C were selected as candidates since the input motion is at bedrock in this site response analysis. No limit on scale factors was enforced, which is a debatable practice. However, match to spectral shape has been identified as the single most important factor in ground motion selection (NEHRP Consultants Joint Venture, 2011). Therefore, some engineering judgment is necessary when the choice has to be made between two candidate records – one with a scale factor close to 1 but a poor match to spectral shape, and another with a high scale factor but a close match to spectral shape. For similar reasons, a wide range of source-to-site distances was permitted after initially attempting to limit the candidates to a range of R = 20–90 km, resulting in an unacceptable match to the target spectrum.

Chapter 5 provides a detailed discussion of measuring match to spectral shape and computing scale factors in amplitude scaling of ground motion records.

The resulting ground motion pairs with scale factors are summarized in Table 2.6.1. Figures 2.6.1 and 2.6.2 depict plots of suite average match to the target spectrum. Refer to Chapter 5 for detailed explanations of ground motion selection and modification procedures.

Figure 2.6.3 is a plot of site amplification versus period, indicating significant amplification of ground motion at long periods and de-amplification at short periods for this soft soil site. The structural engineer is advised to consult with experts in site response before applying such results in the final design of structures, as there are numerous parameters requiring careful consideration in this complex analysis.

TABLE 2.6.1
Ground Motions for Site Response

Earthquake	M_W	Year	Station	V_{S30} m/s	Site Class	R, km	Scale Factor
Cape Mendocino	7.01	1992	Cape Mendo	568	C	0	0.671
Chi-Chi Taiwan	7.62	1999	TCU089	672	C	0	1.038
Amberley NZ	7.85	2016	MOLS	585	C	46	2.175
Amberley NZ	7.85	2016	SEDS	436	C	34	0.513
Chi-Chi Taiwan	7.62	1999	ILA010	573	C	78	5.066
Chi-Chi Taiwan	7.62	1999	TCU067	615	C	3	0.687
Hector Mine	7.13	1999	WW Trout	425	C	63	5.612
Chi-Chi Taiwan	7.62	1999	CHY022	564	C	63	5.279
El Mayor Cucapah	7.20	2010	SIV	483	C	55	14.140
El Mayor Cucapah	7.20	2010	Tres Herm	522	C	95	12.596
Quellon Chile	7.60	2016	LL07	462	C	75	0.915

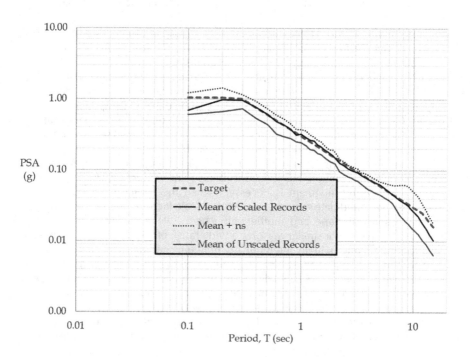

FIGURE 2.6.1   Target and suite mean spectra.

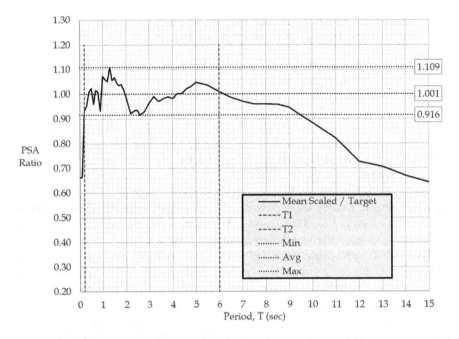

FIGURE 2.6.2   Suite to target PSA ratio.

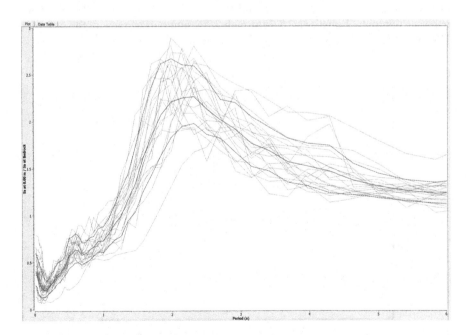

FIGURE 2.6.3   Site response amplification.

2.7 SUBSTITUTE STRUCTURE METHOD FOR INELASTIC RESPONSE SPECTRUM ANALYSIS

The substitute structure method (SSM) of analysis accounts for inelastic behavior in a response spectrum analysis through (a) the incorporation of secant stiffness, K_{EFF}, (b) an increased effective damping, ξ_{EFF}, through hysteretic behavior, and (c) response modification, R_{ξ}, resulting from the increased damping. The method was first proposed by Gulkan and Sozen (Gulkan & Sozen, December, 1974) and was further developed by Priestley (Priestley, Calvi, & Kowalsky, 2007) and others. The method is incorporated into the design provisions for seismic isolation design of bridges in AASHTO (AASHTO, 2014).

SSM analysis may be useful in estimating inelastic response of a structure subjected to seismic loading using only a design response spectrum. Ground motion suites are not required for SSM analysis. The method is likely most appropriate for preliminary design, with final design completed using a nonlinear response history analysis with an appropriately selected and modified suite of ground motion records.

Figure 2.7.1 depicts a simple bilinear load-displacement curve typically used in SSM analysis. The behavior is defined by three parameters: (a) the yield displacement, D_y, (b) the initial stiffness, K_i, and (c) the post-yield stiffness ratio, α. The effective damping is typically expressed as the sum of an initial, elastic component and a hysteretic component, with the hysteretic component dependent on displacement ductility, defined in Equation 2.7-1. Various relationships have been proposed for both the effective damping, ξ_{EFF}, and the response modification, R_{ξ}, due to the increased damping. Equations 2.7-2 through 2.7-8 summarize a few of the proposed relationships.

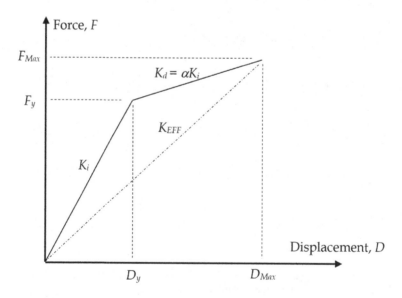

FIGURE 2.7.1 Bilinear load-deflection curve for SSM analysis.

Equation 2.7-2, for use with reinforced concrete bridge substructures, and Equation 2.7-4, for steel frame buildings, were proposed by Priestley (Priestley, Calvi, & Kowalsky, 2007). Equation 2.7-3 is the theoretical value for a bilinear oscillator. Since a system is displaced at the maximum value, D_{Max}, typically for only a small portion of the loading history during strong ground shaking, some fraction of the hysteretic component in Equation 2.7-3 may be appropriate for SSM analysis. The elastic component is frequently taken as a much smaller value than the historically proposed value of 5% used in elastic response spectrum analysis.

Equations 2.7-5 and 2.7-6 have been proposed for use in Euro Code 8 for non-pulse and pulse-type ground motions, respectively. Equation 2.7-7 is the version of response modification included in isolation design for bridges in the United States (AASHTO, 2014). The response modification factor given by Equation 2.7-8 is dependent not only on effective damping but also on significant duration, D_{5-95}, of the ground motion. The model was proposed by Stafford and others (Stafford, Mendis, & Bommer, August, 2008).

SSM analysis requires evaluation of the effective period, T_{EFF}, given by Equation 2.7-9. The maximum response, D_{Max}, is given by Equation 2.7-10. PSA_{TEFF} is the pseudo-spectral acceleration from the design response spectrum at a period equal to T_{EFF}, not the initial period, T_i. Note that the maximum response depends on the effective damping, which depends on the maximum response. Hence, the iterative nature of SSM analysis becomes evident.

$$\mu_D = \frac{D_{Max}}{D_y} \tag{2.7-1}$$

$$\xi_{EFF} = 0.05 + 0.444\frac{\mu_D - 1}{\pi\mu_D} \tag{2.7-2}$$

$$\xi_{EFF} = \xi_{EL} + \frac{2(\mu_D - 1)(1 - \alpha)}{\pi\mu_D(1 + \alpha\mu_D - \alpha)} \tag{2.7-3}$$

$$\xi_{EFF} = 0.05 + 0.577\frac{\mu_D - 1}{\pi\mu_D} \tag{2.7-4}$$

$$R_\xi = \left(\frac{0.10}{0.05 + \xi_{EFF}}\right)^{0.50} \tag{2.7-5}$$

$$R_\xi = \left(\frac{0.07}{0.02 + \xi_{EFF}}\right)^{0.25} \tag{2.7-6}$$

$$R_\xi = \left(\frac{0.05}{\xi_{EFF}}\right)^{0.30} \geq 0.588 \tag{2.7-7}$$

$$R_\xi = 1 - \frac{-0.631 + 0.421 \ln\left(\xi_{EFF}\right) - 0.015 \ln\left(\xi_{EFF}\right)^2}{1 + \exp\left\{\dfrac{-\left[\ln\left(D_{5-95}\right) - 2.047\right]}{0.930}\right\}} \tag{2.7-8}$$

$$T_{EFF} = 2\pi \sqrt{\frac{W}{gK_{EFF}}} = T_i \sqrt{\frac{\mu_D}{1 + \alpha\mu_D - \alpha}} \tag{2.7-9}$$

$$D_{Max} = \left(PSA_{TEFF} \cdot g\right)\left(\frac{T_{EFF}}{2\pi}\right)^2 \cdot R_\xi \tag{2.7-10}$$

The procedure for SSM analysis for a given single-degree-of-freedom system with given initial stiffness, yield strength, mass, and post-yield stiffness ratio is iterative in nature and may be summarized in the following steps.

1. Determine the period, T_i, based on initial stiffness, and the yield displacement, $D_y = F_y/K_i$.
2. Assume a value for displacement ductility demand, μ_D. See Equation 2.7-1.
3. Calculate the effective period, T_{EFF}. See Equation 2.7-9.
4. From the design response spectrum, determine PSA_{TEFF} (g), the pseudo-spectral acceleration at a period equal to T_{EFF}.
5. Calculate the effective damping, ξ_{EFF}, from the appropriate rule for the structure being analyzed. See Equations 2.7-2 through 2.7-4.
6. Calculate the response modification factor, R_ξ, from the appropriate rule for the structure being analyzed. See Equations 2.7-5 through 2.7-8.
7. Calculate the maximum displacement, D_{Max}. See Equation 2.7-10.
8. Calculate the resulting ductility demand, $\mu_D = D_{Max}/D_y$.
9. If the displacement ductility calculated in Step 8 agrees with the assumed value in Step 2, the analysis is complete and the solution has converged. Otherwise, return to Step 3 with a revised ductility demand assumption.

The procedure may be adopted to permit the calculation of the required yield force necessary to limit response to a given ductility demand. This is the basis of the direct displacement-based design method gaining popularity in seismic design.

Example 2.7-1: SSM Analysis of a Structure with Known Properties

A bridge has been modeled as a single-degree-of-freedom system with the following properties and design spectrum control points. Refer to Section 2.4 for the spectral shape definition, given S_{DS} and S_{D1}. Use Equation 2.7-6 for pulse-type ground motion and determine the maximum displacement and force experienced by the system subjected to seismic loading defined by the response spectrum. For this example, use the effective damping rule given in Equation 2.7-3 with only

50% of the hysteretic component assumed in the calculations, with an initial elastic component equal to zero.

$W = 9,896$ kips $\qquad$ $K_i = 2,395$ k/in $\qquad$ $\alpha = 0.02$

$F_y = 3,000$ kips $\qquad$ $S_{DS} = 0.917$ g $\qquad$ $S_{D1} = 0.655$ g

Several iterations performed with spreadsheet calculations were used to establish a converged ductility demand equal to 5.34. The detailed calculations for this final iteration are shown below.

$$T_i = 2\pi\sqrt{\frac{9,896}{386 \times 2,395}} = 0.65 \text{ sec}$$

$$\text{Assume } \mu_D = 5.34$$

$$T_{EFF} = 0.65 \times \sqrt{\frac{5.34}{1+0.02(5.34)-0.02}} = 1.44 \text{ sec}$$

$$\xi_{EFF} = 0.00 + 0.50 \times \frac{2(5.34-1)(1-0.02)}{\pi(5.34)(1+0.02\times5.34-0.02)} = 0.233$$

$$R_\xi = \left(\frac{0.07}{0.02+0.233}\right)^{0.25} = 0.725$$

For PSA_{TEFF}, note that $T_S = S_{D1}/S_{DS} = 0.655/0.917 = 0.714$ sec and $T_o = 0.2T_S = 0.143$ sec. With an effective period $T_{EFF} = 1.44$ sec, PSA_{TEFF} may be computed.

$$PSA_{TEFF} = \frac{0.655}{1.44} = 0.455 \text{ g}$$

$$D_{Max} = (0.455 \cdot 386)\left(\frac{1.44}{2\pi}\right)^2 \cdot 0.725 = 6.69 \text{ inches}$$

$$D_y = \frac{3,000}{2,395} = 1.25 \text{ inches}$$

$$\mu_D = \frac{6.69}{1.25} = 5.34 = \text{Assumed Value} \rightarrow \text{Convergence reached}$$

$$F_{Max} = F_y + K_d(D_{Max} - D_y) = 3,000 + 0.02(2,395)(6.69 - 1.25) = 3,260 \text{ kips}$$

GM-Parameters		R_d-Parameters		SSM-Parameters	
A_S (g) =:	0.5529	T* (sec) =:	0.893	Hysteresis :	Bi-Linear
S_{DS} (g) =:	0.9170	use T* (sec) =:	0.893	ξ_0 :	0
S_{DI} (g) =:	0.6550	Inelastic-Parameters		ξ_{EFF} Rule :	Pure Bi-Linear
T_L (sec) =:	12	α =:	0.02	R_ξ Rule :	EC8-Pulse
D_{5-95} (sec) =:	35	Assumed μ =:	6.000	κ :	0.50
T_S (sec) =:	0.714	Calculated μ =:	6.000	ξ_{MAX} :	0.30
T* (sec) =:	0.893	Calculated D =:	7.030	ξ_{NLRHA} :	0.00000
T_O (sec) =:	0.143	T_{BFF}/T_i =:	2.335	Tolerance :	0.001
W, kips	9896	Dy, inches	1.172	SA(T)	0.9170
k_i, kips/inch	2395	Assumed D, inches	7.030	SA(T_{EFF})	0.4315
F_y, kips	2806	T, sec	0.6500	T_{EFF}, sec	1.5181
F_{MAX}, kips	3086.69	F_{MAX}, kips	3086.69	k_{EFF}, k/in	439.084

FIGURE 2.7.2 Spreadsheet SSM analysis for DDBD.

Example 2.7-2: Direct Displacement-Based Design Using SSM Analysis

A single-story building has the following properties and design spectrum control points. Determine the yield force necessary to limit the displacement ductility to no more than 6. Use Equation 2.7-6 for pulse-type ground motion and determine the maximum displacement and force experienced by the system subjected to seismic loading defined by the response spectrum. For this example, use the effective damping rule given in Equation 2.7-3 with only 50% of the hysteretic component assumed in the calculations, with an initial elastic component equal to zero.

W = 9,896 kips K_i = 2,395 k/in α = 0.02
S_{DS} = 0.917 g S_{DI} = 0.655 g

Spreadsheet solution was used and the converged solution is shown in Figure 2.7.2. The required yield force is found to be 2,806 kips.

3 Building Design

The following material includes topics not typically covered in undergraduate structural engineering curricula. Nonetheless, the material frequently arises in practice early in engineering careers. A basic understanding of structural steel design in accordance with AISC 360-16 is a prerequisite for a full appreciation of the material presented.

3.1 COMPOSITE BEAM DESIGN

The design of composite beams for buildings will be discussed for both rolled steel shapes and welded plate girders. While welded plate girders are not as common as rolled shapes for building design, the need for such girders does arise. Open areas requiring long spans may necessitate the use of a plate girder in a building. Rigid frames with tapered webs are sometimes found in industrial facilities, and plate girders are needed to accomplish the desired structure. Plate girders are very common in bridge structures.

AISC 360-16 (American Institute of Steel Construction, 2016) Chapter I contains basic design criteria for composite beams used in buildings. Deflection criteria are not covered in AISC. These may include requirements for both superimposed loading after the building is in service, and construction requirements prior to deck strength being fully achieved. For example, the owner may require that deflection due to superimposed live load be no more than $L/360$. This would be the criteria to be satisfied using the composite properties to determine actual deflection. Similarly, the engineer may require that the deflection due to wet concrete plus a construction live load allowance be no greater than $L/240$. This would be a requirement on the properties of the beam alone.

For flexural design of beams, the engineer needs to determine the number and size of shear studs required to transmit horizontal shear between the deck and the beam. This load transfer between a steel beam and the concrete slab is accomplished by designing for a horizontal shear force equal to the smaller of Equations 3.1-1 and 3.1-2, for full composite action.

$$V' = 0.85 f'_c A_c \tag{3.1-1}$$

$$V' = F_y A_s \tag{3.1-2}$$

To avoid problems related to the deformation capacity of shear studs, the Commentary to Chapter I of AISC 360-16 notes that either a nonlinear deformation capacity analysis should be performed or else one or more of the following should be satisfied.

1. The beam span does not exceed 30 ft.
2. The degree of composite action is at least 50%.
3. The average nominal shear stud capacity is at least 16 kips/ft.

According to the Commentary, if any of the above three criteria are satisfied, then deformation capacity of the shear studs need not be verified by nonlinear analysis.

Shear stud capacity is given by AISC 360-16, Section I8.2a, with Table 3-21 of the AISC Steel Construction Manual providing the engineer with typical values for the most commonly encountered situations in practice.

If a shear stud arrangement is developed that provides a lower capacity than either of Equations 3.1-1 and 3.1-2, then the section is said to be partially composite. Partial composite design is often a valid design option.

The horizontal shear force is to be carried by deck-beam interface shear studs between the points of zero and maximum moment. In the 28-ft long beam represented by the moment diagram of Figure 3.1.1, the calculated shear stud capacity must be satisfied over one-half of the beam span. In the 21-ft long girder represented by the moment diagram of Figure 3.1.2, the requirement must be satisfied over one-third of the beam span.

Example 3.1-1: Composite Beam Analysis

A simply supported, uniformly loaded W21x44 beam spans 30 ft and supports a tributary width of 10 ft. The beam is not an edge beam and has deck spans on both sides. Steel is A992 and concrete strength is 4,000 psi. The deck thickness is 4.5 inches.

(a) Determine the effective flange width of the composite beam.
(b) Determine the nominal and design moment resistances with full composite action.

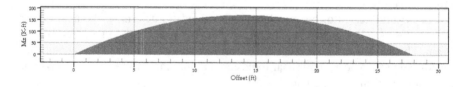

FIGURE 3.1.1   Moment diagram – simply supported uniformly loaded beam.

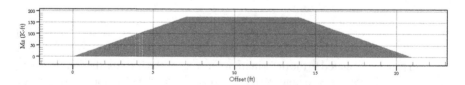

FIGURE 3.1.2   Moment diagram – girder loaded at 1/3 points.

(c) Determine the number of ¾-inch Type B shear studs required to achieve full composite action (F_u = 65 ksi for Type B studs).
(d) How much floor live load may be placed without exceeding the moment resistance?
(e) Re-compute the design moment strength and live load if 40 ¾-inch Type B shear studs are spaced uniformly over the entire beam length.
(f) Compare the fully composite design moment strength to the design moment strength of the beam alone. Assume the slab fully braces the compression flange.

(a) Effective flange width (AISC 360-16, I3.1a)

$$b_{e1} = \frac{30}{8} \times 2 = 7.5\,\text{ft} = 90\,\text{inches} \leftarrow \text{Controls}$$

$$b_{e2} = \frac{1}{2} \times 10\,\text{ft} \times 2 = 10\,\text{ft} = 120\,\text{inches}$$

$$\Rightarrow b_e = 90\,\text{inches}$$

(b) ϕM_n with full composite action (W21x44 → A_s = 13.0 in², A992 steel → F_y = 50 ksi)

$$V_1' = 0.85 \times 4 \times 90 \times 4.5 = 1{,}377\,\text{kips}$$

$$V_2' = 13.0 \times 50 = 650\,\text{kips} \leftarrow \text{Controls}$$

$$V' = 650\,\text{kips}$$

Since the steel section controls the maximum force that is to be carried across the deck-beam interface, the deck is only partially in compression. Determine the depth of the compressive stress block, a, and the distance from the top of the beam to the compressive force, Y2 (page 3-13 of the AISC Steel Construction Manual.

$$a = \frac{650}{0.85 \times 4 \times 90} = 2.12\,\text{inches}$$

$$Y2 = 4.5 - \frac{2.12}{2} = 3.44\,\text{inches}$$

The distance from the top of the beam to the center of the tensile force in the beam, $Y3$, will be simply one-half of the beam depth, in this case. The nominal moment resistance, M_n, is the force multiplied by the distance between the compressive and tensile centroids. The appropriate resistance factor from AISC 360-16 is 0.90.

$$Y3 = \frac{20.7}{2} = 10.35 \text{ inches}$$

$$\phi M_n = 0.90(650)(3.44 + 10.35)/12 = 672 \text{ ft} \cdot \text{kips}$$

(c) Shear stud design (3/4-inch diameter, F_u = 65 ksi, AISC 360-16, I8.2a)

$$A_{sa} = \frac{\pi}{4}(0.75)^2 = 0.442 \text{ in}^2$$

$$E_c = 145^{1.5}\sqrt{4} = 3,492 \text{ ksi}$$

$$R_g = 1.0, \text{weld directly to beam}$$

$$R_p = 0.75, \text{weld directly to beam}$$

$$Q_n = 0.5(0.442)\sqrt{4 \times 3,492} = 26.1 \text{ kips/stud}$$

$$Q_n = 0.75 \times 1.0 \times 0.442 \times 65 = 21.5 \text{ kips/stud} \leftarrow \text{Controls}$$

$$n \geq \frac{650}{21.5} = 30.2 \text{ studs between mid-span and end}$$

Use 29 spaces at 6 inches on center, each side of mid-span (30 studs ≈ 30.2 studs, say ok)

(d) Maximum permissible live load

Allow for a 5% contingency on self-weight to account for connections. The dead load for this example is simply the beam self-weight plus the concrete deck weight.

$$w_{deck} = \frac{4.5}{12} \times 0.150 \times 10' = 0.562 \text{ klf}$$

$$W_D = 1.05 \times 0.044 + 0.562 = 0.609 \text{ klf}$$

$$M_u = \frac{w_u L^2}{8} = \frac{w_u \times 30^2}{8} \leq 672 \Rightarrow w_u \leq 5.97 \text{ klf}$$

$$1.2(0.609) + 1.6(w_L) \leq 5.97 \Rightarrow w_L \leq 3.28 \text{ klf}$$

$$q_L \leq \frac{3.28}{10'} = 0.33 \text{ ksf} = 330 \text{ psf}$$

(e) 40 studs over the entire beam length

This represents 20 studs between the points of zero (beam end) and maximum (mid-span) moments. This is less than the 30 required for full composite action, so the beam is partially composite.

$$V' = 20 \text{ studs} \times 21.5\frac{\text{kips}}{\text{stud}} = 430 \text{ kips}$$

$$a = \frac{430}{0.85 \times 4 \times 90} = 1.40 \text{ inches}$$

$$Y2 = 4.5 - \frac{1.40}{2} = 3.80 \text{ inches}$$

From Table 3-19 of the AISC Steel Construction Manual (which may be used since $F_y = 50$ ksi for the beam):

$$\phi M_n = 615 + (631 - 615)(0.3 / 0.5) = 625 \text{ ft} \cdot \text{kips}$$

(f) Composite versus non-composite flexural strength

From Table 3-2 of the AISC Steel Construction Manual, for $L_b = 0$:

$$\phi M_n = \phi M_p = 358 \text{ ft} \cdot \text{kips, for the W 21X 44 noncomposite}$$

The fully composite beam is seen to have a flexural resistance 88% greater than that of the non-composite beam.

Section 3 in the AISC Steel Construction Manual contains tables which could be used to shorten the calculations in the preceding example since the beam is A992 steel.

Example 3.1-2: Composite Plate Girder Analysis

The plate girder shown in Figure 3.1.3 has concentrated dead and live loads at mid-span and a uniformly distributed self-weight. The plate girder is simply supported and is laterally braced at quarter-points and ends. The steel cross section is shown in Figure 3.1.4. The deflection of the girder is to be limited to $L/600$ during construction. The in-service girder will be fully composite with an 8-inch thick concrete deck having an effective width equal to 12 ft 6 inches (1/4 of the span, based on AISC 360-16, Section I3.1a).

The plate girder is ASTM A572 Grade 50 material. The flange-to-web welds will be designed for both continuous and intermittent fillet welds using 70 ksi electrodes. The anticipated concentrated loads during construction before the deck have attained the specified 28-day compressive strength of 6,000 psi, but after the quarter-point bracing has been installed, are $P_D = 240$ kips and $P_L = 160$ kips. The in-service loads are $P_D = 400$ kips and $P_L = 350$ kips.

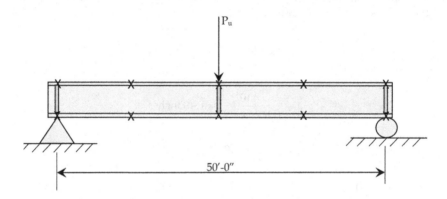

FIGURE 3.1.3   Composite plate girder elevation.

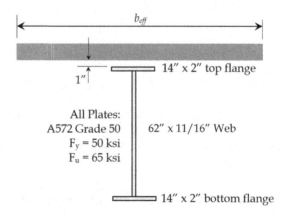

FIGURE 3.1.4   Composite plate girder cross section.

For the non-composite girder, before the deck strength has been attained, the girder cross section properties are needed. The factored loads for the construction condition are needed as well. In addition to the typical properties needed, AISC 360-16 Section F2 is used to determine the limiting unbraced lengths, L_p and L_r.

$$P_u = 1.2 \times 240 + 1.6 \times 160 = 544 \text{ kips}$$

$$P_s = 240 + 160 = 400 \text{ kips}$$

$$A_g = 62 \times 0.6875 + 14 \times 2 \times 2 = 98.625 \text{ in}^2$$

$$w_s = 0.490 \times \frac{98.625}{144} = 0.336 \text{ klf}$$

$$w_u = 1.2 \times 0.336 = 0.403 \text{ klf}$$

$$I_x = \frac{2 \times 14 \times 2^3}{12} + 2 \times (14 \times 2) \times 32^2 + \frac{0.6875 \times 62^3}{12} = 71,017 \text{ in}^4$$

$$S_x = \frac{71,017}{33} = 2,152 \text{ in}^3$$

$$Z_x = 2 \times (31 \times 0.6875) \times 15.5 + 2(14 \times 2) \times 32 = 2,453 \text{ in}^3$$

$$M_p = 50 \times 2,453 = 122,650 \text{ in} \cdot \text{k} = 10,221 \text{ ft} \cdot \text{k}$$

$$I_y = \frac{62(0.6875)^3}{12} + \frac{2 \times 2(14)^3}{12} = 916 \text{ in}^4$$

$$r_y = \sqrt{\frac{916}{98.625}} = 3.05 \text{ inches}$$

$$L_p = 1.76(3.05)\sqrt{\frac{29,000}{50}} = 129 \text{ inches} = 10.8 \text{ ft}$$

$$h_o = 64 \text{ inches}$$

$$J \cong \sum \frac{bt^3}{3} = \frac{2 \times 14 \times 2^3}{3} + \frac{62 \times 0.6875^3}{3} = 81.4 \text{ in}^4$$

$$r_{TS} = \sqrt{\frac{I_y h_o}{2S_x}} = \sqrt{\frac{916 \times 64}{2 \times 2,152}} = 3.69 \text{ inches}$$

$$\frac{Jc}{S_x h_o} = \frac{81.4 \times 1}{2,152 \times 64} = \frac{1}{1,692}$$

$$6.76 \left(\frac{0.7F_y}{E} \right)^2 = 6.76 \left(\frac{0.7 \times 50}{29,000} \right)^2 = \frac{1}{101,558}$$

$$L_r = 1.95(3.69 \text{ inches}) \frac{29,000}{0.7 \times 50} \sqrt{\frac{1}{1,692} + \sqrt{\left(\frac{1}{1,692} \right)^2 + \frac{1}{101,558}}}$$

$$= 367 \text{ inches} = 30.5 \text{ ft}$$

$$10.8 \text{ ft} < L_b = 12.5 \text{ ft} < 30.5 \text{ ft}$$

Since the actual unbraced length, L_b, is between L_p and L_r, the lateral torsional buckling (LTB) modification factor, C_b, is needed in order to determine the design flexural resistance.

$$M_u = \frac{544 \times 50}{4} + \frac{0.403(50)^2}{8} = 6,800 + 126 = 6,926 \text{ ft} \cdot k = M_{max}$$

For the interior unbraced segment adjacent to mid-span, where M_{max} occurs, point "A" in the Equation for C_b is at $x = 15.625$ ft, point "B" is at $x = 18.75$ ft, and point "C" is at $x = 21.875$ ft. The LTB modification factor may be determined using AISC 360-16 Equation F1-1.

$$M_A = 6,800 \frac{15.625}{25} + \frac{0.403 \times 15.625}{2}(50 - 15.625) = 4,358 \text{ ft} \cdot k$$

$$M_B = 6,800 \frac{18.75}{25} + \frac{0.403 \times 18.75}{2}(50 - 18.75) = 5,218 \text{ ft} \cdot k$$

$$M_A = 6,800 \frac{21.875}{25} + \frac{0.403 \times 21.875}{2}(50 - 21.875) = 6,074 \text{ ft} \cdot k$$

$$C_b = \frac{12.5(6,926)}{2.5(6,926)+3(4,358)+4(5,218)+3(6,074)} = 1.246$$

$$M_n = 1.246\left[10,221-(10,221-6,277)\left(\frac{12.5-10.8}{30.5-10.8}\right)\right] = 12,311 > M_p = 10,221$$

$$\rightarrow \phi M_n = \phi M_p = 0.90 \times 10,221 = 9,199 \text{ ft} \cdot \text{k} > M_u = 6,926 \text{ ft} \cdot \text{k, OK}$$

For the design shear resistance of the unstiffened girder, AISC 360-16 Section G2 provides the necessary equations.

$$k_v = 5.34$$

$$\frac{h}{t_w} = 90.2 > 1.10\sqrt{\frac{5.34 \times 29,000}{50}} = 61.2$$

$$\rightarrow C_{v1} = \frac{61.2}{90.2} = 0.679$$

$$\phi V_n = 0.90\left[0.6 \times 50 \times 66 \times 0.6875 \times 0.679\right] = 832 \text{ kips}$$

$$V_u = \frac{544}{2} + 0.403 \times 25 = 282 \text{ kips} < 832 \text{ kips, OK}$$

For the deflections under service loads, equations from Section 3 of AISC 360-16 may be used.

$$\Delta = \frac{400(50 \times 12)^3}{48 \times 29,000 \times 71,017} + \frac{5(0.336/12)(50 \times 12)^4}{384 \times 29,000 \times 71,017}$$

$$= 0.874 \text{ inch} + 0.023 \text{ inch} = 0.897 \text{ ft}$$

$$\Delta_{limit} = \frac{50 \times 12}{600} = 1.00 \text{ inch, OK}$$

AISC 360-16, Section F13, prescribes proportioning limits for plate girders. These limits should always be checked. For this doubly symmetric cross section, I_{yc}/I_{yt} = 1 and easily satisfies F13.2. The web slenderness, h/t_w= 90.2, also satisfies the requirements easily, as identified in F13.3. Should a field splice be required for the girder, F13.1 may be used to determine the maximum number of holes that may be placed in the tension flange without reduction in flexural resistance. For this example, use 1-inch diameter bolts in standard holes having a diameter of 1.125 inches in accordance with AISC 360-16 Table J3.3. For net area calculation, AISC 360-16 Section B4.3b requires that the diameter of the hole be equal to 1/16 inch greater than the standard hole diameter to account for possible damage to the hole in the drilling/punching process.

$$\frac{F_y}{F_u} = \frac{50}{65} = 0.769 < 0.80 \rightarrow Y_t = 1.0$$

$$A_{fn} = 2\left[14 - n_{holes} \times 1.1875\right] = 28 - 2.375n_{holes}$$

$$A_{fn} \geq \frac{Y_t F_y}{F_u} A_{fg} = 0.769 \times 28 = 21.53 \text{ in}^2$$

$$28 - 2.375n_{holes} \geq 21.53 \rightarrow n_{holes} \leq 2.72 \rightarrow \text{use 2 bolts per row}$$

Should the condition not be satisfied, the reduction in flexural resistance might be significant. Suppose the engineer decided to use four bolts per row in a bolted field splice for the girder.

$$A_{fn} = 2\left[14 - 4 \times 1.1875\right] = 18.52 \text{ in}^2$$

$$\phi M_n = 0.90 \times F_u \times \frac{A_{fn}}{A_{fg}} \times S_x$$

$$= 0.90 \times 65 \times \frac{18.52}{28.00} \times 2,152$$

$$= 83,269 \text{ in} \cdot \text{k} = 6,939 \text{ ft} \cdot \text{k}$$

While the design flexural resistance is still greater than the required resistance of 6,926 ft·k, it is significantly less than the design resistance of 9,199 ft·k, had only two holes per row in the flange been specified.

For the composite girder, AISC 360-16 Section I3.1b permits positive flexural resistance to be based on the plastic force distribution on the cross section if the web satisfies depth-to-thickness criteria. Otherwise, positive flexural resistance is to be based on the superposition of elastic stresses at the various stages of construction.

$$\frac{h}{t_w} = \frac{62}{0.6875} = 90.2 < 3.76\sqrt{\frac{29,000}{50}} = 90.6 \rightarrow \text{use plastic analysis}$$

Determine the location of the plastic neutral axis (PNA). This requires computation of the force in each of the elements assuming steel is fully yielded and concrete stress is 0.85 f'_c.

$$P_{deck} = 0.85 \times 6 \times 150 \times 8 = 6,120 \text{ kips}$$

$$P_{girder} = 50 \times 98.625 = 4,931 \text{ kips}$$

Since the tension in the fully yielded girder is less than the force in the full compressed deck, the PNA lies in the deck.

By Section I3.2d of AISC 360-16, the force to be transferred between the deck and the girder for fully composite behavior is controlled by the steel. The depth of deck in compression and the distance between the compressive and tensile components of the flexural couple acting on the cross section can be determined. See also page 3-13 of the 15th edition of the AISC Steel Construction Manual. For calculation of M_p, the 1-inch gap between the top of the girder and the bottom of the deck will be ignored in this example, in consideration of construction tolerances. This conservative assumption is not necessary. If the 1-inch gap is included, a slightly higher plastic moment would be obtained, and it is not outside the realm of reason to expect the 1-inch gap to be maintained.

$$V' = 4,931 \text{ kips}$$

$$a = \frac{4,931}{0.85 \times 6 \times 150} = 6.45 \text{ inches}$$

$Y_{con} = 1 + 8 = 9$ inches, but ignore 1 inch gap and use $Y_{con} = 8$ inches

$$Y2 = 8 - \frac{6.45}{2} = 4.78 \text{ inches}$$

The distance from the top of the beam to the tensile force in the girder is simply one-half of the girder depth in this case, with the girder entirely in tension. Y2 is the distance from the top of the steel to the center of the deck compressive force.

$$M_p = 4,931\left(4.78 + \frac{66}{2}\right) = 186,293 \text{ in·k} = 15,524 \text{ ft·k}$$

$$\phi M_n = 0.90 \times 15,524 = 13,973 \text{ ft·k}$$

$$P_u = 1.2 \times 400 + 1.6 \times 350 = 1,040 \text{ kips}$$

$$M_u = \frac{1,040 \times 50}{4} + \frac{0.403(50)^2}{8} = 13,000 + 126 = 13,126 \text{ ft} \cdot \text{k} < 13,973 \text{ ft} \cdot \text{k}, \text{OK}$$

$$V_u = \frac{1,040}{2} + 0.403 \times 25 = 530 \text{ kips} < 832 \text{ kips}, \text{OK}$$

Finally, determine the number of shear studs required between the points of zero and maximum moment (between each end and mid-span, in this example). AISC 360-16 Section I8.2a provides the design resistance equations for the metal studs welded directly to the girder. With no metal deck specified in this example, $R_g = 1.0$ and $R_p = 0.75$. For 1-inch diameter studs, $A_{sa} = 0.785 \text{ in}^2$.

$$0.50\sqrt{6 \times 4,458} = 81.77 > R_g R_p F_u = 1.0 \times 0.75 \times 65 = 48.75$$

$$Q_n = 48.75 \times 0.785 = 38.3 \text{ kips per stud}$$

$$n_{reqd} = \frac{4,931}{38.3} = 129 \text{ studs}$$

Use three studs per row across the flange width. Use a 6-inch spacing between rows. This equates to 48 rows between the end and mid-span.

$$n_{actual} = 3 \times 48 = 144 \text{ studs between ends and mid-span}$$

3.2   DIRECT ANALYSIS AND DESIGN

AISC 360-16 addresses specific stability criteria for steel building structures. These criteria may be satisfied in several ways, the most preferred (author's opinion) of which is termed the direct analysis method (DAM).

The direct analysis method may be used for any structural steel building. The equivalent length method and the first-order method are limited in their applicability by AISC 360-16 Appendix 7 (see article 7.2 for the effective length method and article 7.3 for the first-order method). The DAM is the focus of the material presented here.

The DAM of AISC 360-16 includes each of the following (see AISC 360-16 article C2):

1. The analysis must consider flexural, shear, and axial deformations. As noted in the Commentary to C2.1, "consider" does not necessarily equate to "include."
2. The analysis must be a second-order analysis considering both $P\delta$ and $P\Delta$ effects. Approximate second-order analysis techniques found in AISC 360-16 Appendix 8 are permitted.
3. The analysis must include either (a) explicit modeling of imperfections or (b) notional loads to estimate the effects of imperfections.

4. The analysis must include stiffness reduction as a means of incorporating residual stress effects. Two reductions are specified: (a) a factor of 0.8 on all stiffness values contributing to the lateral stability of the structure and (b) a factor, τ_b, applied to flexural stiffness of all members contributing to lateral stability of the structure.
5. Member effective lengths, L_c, are taken equal to the actual member length unless smaller values can be justified by analysis. This is advantageous in that, contrasted with the effective length method, which requires determination of appropriate K-values, $K = 1$ for all members in the DAM.

Given that the DAM requires a second-order analysis, it is clearly not permissible to analyze load cases separately and then combine the effects. Each load combination must be analyzed separately in any second-order, or otherwise nonlinear, analysis.

Example 3.2-1: Direct Analysis Method Frame Example

Figure 3.2.1 is a plane frame with applied dead, live, and wind loads. The frame has been studied by multiple researchers. Steel is Grade 50. Assess the adequacy

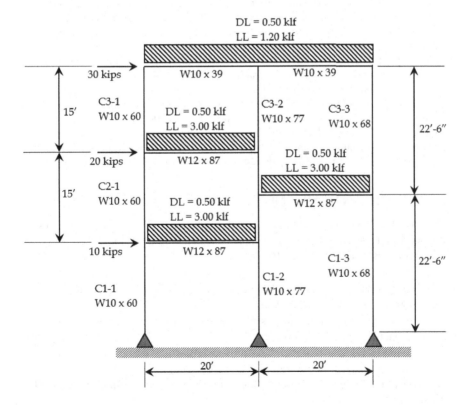

FIGURE 3.2.1   Plane frame for direct analysis method solution.

TABLE 3.2.1

Summary of Direct Analysis Method Example Design Ratios

| Member | Design Ratio for Various Analyses | | |
	First Order	Second Order	Direct
C1-1	1.123	1.087	1.156
C2-1	0.312	0.312	0.320
C3-1	0.307	0.308	0.310
C1-2	1.249	1.406	1.320
C2-2a	0.387	0.391	0.396
C2-2b	0.506	0.523	0.530
C3-2	0.324	0.330	0.334
C1-3	0.800	0.882	0.916
C3-3	0.340	0.340	0.340
Floor beams	0.938	1.055	1.103
Roof beams	0.711	0.720	0.747

of the design for the load combination – 1.2D + 1.6L + 0.8W – using (a) first-order analysis, (b) second-order analysis, and (c) the AISC direct analysis method.

An IES VisualAnalysis model of the plane frame was used to determine the design ratios for the members using each of the three specified analysis strategies. Recall that the DAM requires not only a second-order analysis but also consideration of reduced stiffness and notional loads. Table 3.2.1 summarizes the results and makes clear the importance of including all required factors in the design of modern steel structures – slender, or non-symmetric steel structures, in particular. The design ratio is defined as the required resistance divided by the available resistance. Hence, a value close to, but less than, 1.0 indicates a safe and economical design. A value much less than 1.0 represents an overdesign. A value greater than 1.0 represents a failure to satisfy design requirements.

3.3   PLASTIC ANALYSIS AND DESIGN

AISC 360-16 Appendix 1 permits design by inelastic analysis in certain cases for steel building members. Four criteria must be satisfied in order for inelastic analysis to be permissible (see AISC 360-16 A1.3.2).

1. The specified minimum yield strength, F_y, of members subject to plastic hinging may not exceed 65 ksi.
2. The cross section of members subject to plastic hinging must qualify not as compact but as "ductile-compact."
3. The unbraced length of members subject to plastic hinging is limited not to L_p as is the case for conventional design but to L_{pd}.
4. The axial force in members subject to plastic hinging must not exceed $0.75F_yA_g$.

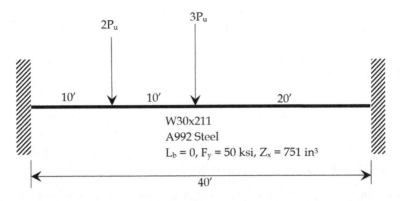

FIGURE 3.3.1   Fixed-fixed beam for inelastic analysis.

TABLE 3.3.1
Progressive Collapse of Fixed-Fixed Beam

Load Factor	Criterion	Element and Location
105.0	First yield	Beam B1, left end
134.0	First yield	Beam B3, right end
144.0	First yield	Beam B2, right end
144	First yield	Beam B3, left end
152.5	Plastic hinge	Beam B1, left end
155.0	Plastic hinge	Beam B3, left end
155.5	Plastic hinge	Beam B2, right end
155.5	Plastic hinge	Beam B3, right end
158.2	Fracture	Beam B1, left end

Before commencing with an example, it is beneficial to recall, from Chapter 2, a few of the different types of structural analysis available to structural engineers.

- First-Order Elastic Analysis – neglects all nonlinearities; good indicator for service load conditions
- Second-Order Elastic Analysis – $P\delta$ and $P\Delta$; no material nonlinearity
- First-Order Inelastic Analysis – equilibrium on undeformed structure; material nonlinearity included
- Second-Order Inelastic Analysis – equilibrium on deformed structure; material nonlinearity included

Example 3.3-1: Inelastic Analysis

Various hand calculation methods for inelastic analysis are presented in the literature (Bruneau, Uang, & Sabelli, 2011). For this example, a computer-aided solution will be obtained using SeismoStruct 2020 (Seismosoft, 2020). Figure 3.3.1 shows a fixed-fixed beam with transverse loads as shown. The material is A992 steel.

Of the possible mechanisms, that with plastic hinges at the ends and at mid-span is found to control using hand calculations. Calculations in Appendix B show that failure occurs when $P_u = 156.5$ kips.

The SeismoStruct model for this beam indicates a nominal load corresponding to $P_u = 155.5$ kips, very close to the hand-calculated load, under progressive collapse as shown in the development of yielding and hinges at the ends and at mid-span. Table 3.3.1 summarizes the progressive collapse.

3.4 LATERAL FORCE RESISTING SYSTEMS

For civil engineering graduates with a structures focus, a basic knowledge of lateral force resisting systems (LFRS) used for buildings is beneficial. ASCE 7-16 and AISC 341-16 (American Institute of Steel Construction, 2016) are two crucial documents related to LFRS criteria for steel buildings. For steel moment frames, AISC 358-16 (American Institute of Steel Construction, 2016) is crucial as well.

ACI 318 for buildings and ACI 349 for safety-related nuclear structures provide design requirements for concrete structures. TMS 402 provides design requirements for masonry structures.

Lateral force resisting systems for concrete and masonry construction include the following, among many others:

- Moment frames – ordinary and special
- Concrete shear walls – ordinary and special
- Reinforced masonry shear walls – ordinary, intermediate, and special
- Plain masonry shear walls

Lateral force resisting systems for steel buildings include the following, among many others:

- Moment frames – ordinary, intermediate, and special
- Concentrically braced frames – ordinary, intermediate, and special
- Eccentrically braced frames
- Buckling-restrained braced frames
- Steel plate shear walls

Each of these steel systems is covered in detail in AISC 341-16. AISC 358-16 contains detailed design provisions for connections in steel intermediate and special moment frames. Modern seismic criteria include so-called capacity design principles wherein a ductile "fuse" is selected and designed for seismic forces reduced to account for inelastic behavior. All other elements are designed for the maximum force that the fuse is capable of delivering to the structural system, the so-called capacity-limited loading. This is the essence of capacity design and ensures that ductile, inelastic behavior is limited to the selected fuse, with all other elements remaining essentially elastic.

For moment frames, the fuse elements are plastic hinges, preferably in the beams and preferably away from the face of the columns. For concentric braced frames, the fuse elements are the braces, yielding in tension and buckling in compression. For eccentric braced frames, the fuse elements are the "links" – short segments subjected to high shear. These links may be located at the ends of beams, within beams, or in columns. Two EBF configurations, with the links identified, are shown in Figure 3.4.1, based on AISC 341-16.

The Split-K EBF (Figure 3.4.1a) offers several advantages over other configurations. These advantages include:

1. Approximate equations for link shear demand are extremely accurate, in general.
2. The link end moments are typically very close to being equal in magnitude, as assumed in the typical analysis.
3. The Split-K requires no beam-to-column connection.
4. The Split-K possesses symmetry.
5. Axial load is typically very small for links in the Split-K EBF.

The mechanics of the Split-K EBF behavior may be summarized from AISC 341-16 as follows, with reference to Figure 3.4.2 (American Institute of Steel Construction, 2016). The shear in a given link may be estimated from Equation 3.4-1. The plastic displacement capacity of an EBF story may be estimated using Equation 3.4-2.

$$V_{Link} = \frac{h}{L} \cdot V_{Story} \qquad (3.4\text{-}1)$$

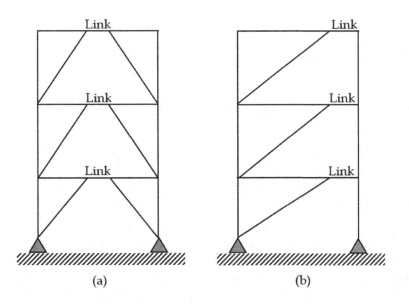

(a)   (b)

FIGURE 3.4.1 EBF configurations (ASCE 341-16).

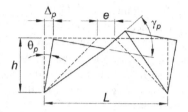

FIGURE 3.4.2   Split-K eccentric braced frame parameters (AISC 341-16).

$$\Delta_p = \frac{h}{L} \cdot \gamma_p e \tag{3.4-2}$$

$$V_P = \begin{cases} 0.60F_y A_{lw}, \left(\dfrac{P_u}{P_y} \le 0.15 \right) \\[4mm] 0.60F_y A_{lw} \sqrt{1 - \left(\dfrac{P_u}{P_y} \right)^2}, \left(\dfrac{P_u}{P_y} > 0.15 \right) \end{cases} \tag{3.4-3}$$

$$A_{lw} = \begin{cases} \left(d - 2t_f \right) t_w, & \text{I-shaped links} \\[2mm] 2t_w \left(d - 2t_f \right), & \text{box-shaped links} \end{cases} \tag{3.4-4}$$

$$M_P = \begin{cases} F_y Z, \left(\dfrac{P_u}{P_y} \le 0.15 \right) \\[4mm] F_y Z \left(\dfrac{1 - P_u / P_y}{0.85} \right), \left(\dfrac{P_u}{P_y} > 0.15 \right) \end{cases} \tag{3.4-5}$$

$$V_n = \text{smaller of} \begin{cases} V_P \\[2mm] 2M_P \Big/ e \end{cases} \tag{3.4-6}$$

$$\text{Link Class} = \begin{cases} \text{Shear,} \ e \le 1.6 M_P / V_P \\ \text{Intermediate,} \ 1.6 M_P / V_P < e \le 2.6 M_P / V_P \\ \text{Flexural,} \ e > 2.6 M_P / V_P \end{cases} \tag{3.4-7}$$

Design shear resistance for the link, ϕV_n, is equal to $0.90V_n$.

Equation 3.4-2 requires the plastic rotation capacity, γ_p. Links are classified as either shear links or flexural links or intermediate links in AISC 341-16. The plastic rotation capacity depends on the link classification. For shear links, $\gamma_p = 0.08$ rad. For flexural links, $\gamma_p = 0.02$ rad. For intermediate links, linear interpolation should be used.

The capacity-limited load, E_{cl}, for EBFs is determined in accordance with AISC 341-16, Section F4.3. The force distribution that produces link shears equals to the adjusted link shear strength ($1.25R_yV_n$ for I-shaped links, $1.40\,R_yV_n$ for box-shaped links) in all levels, and the subsequent response in all braces, beams, connections, and columns defines the capacity-limited condition. It is permitted that design of the beams outside the links be designed for 88% of the capacity-limited loading. Note that, by definition, any software used to evaluate steel design requirements will indicate link failure under the E_{cl} loading. A broad summary of EBF design includes the following steps.

(a) Design the links at each level for the governing load combinations. An example from ASCE 7-16 is $U = 1.2D + 0.5L + 0.2S + 1.0E$.
(b) Determine the capacity-limited load distribution, E_{cl}.
(c) Design beams outside the links for the governing load combination, substituting $0.88E_{cl}$ for E in the combination.
(d) Design braces, connections, and columns for the governing load combination, substituting E_{cl} for E in the combination.
(e) Check drift limits for the designed EBF for the governing load combination (using E, not E_{cl}) remembering that elastic displacements from code-specified lateral force distributions must be amplified (by C_d, for example, in ASCE 7-16) to estimate inelastic displacements.
(f) If drift limits are not satisfied, increase non-link member sizes (most effectively, the braces) until drift limits are satisfied.

Example 3.4-1: Alternative Eccentric Braced Frame Configurations

Consider two models, one for a Split-K EBF (Figure 3.4.3) and another for an alternative configuration (Figure 3.4.4). Figure 3.4.5 shows the axial force, shear, and moment in *Link-01* of the Split-K EBF. Figure 3.4.6 shows the same parameters for the alternate configuration.

The axial force in the Split-K link is zero and the moments at the ends are very close to equal in magnitude, opposite in direction.

While small, the axial force in the link on the alternate configuration is not zero and the end moments are far from equal in magnitude.

Example 3.4-2: Alternating Split-K Eccentric Braced Frame Configurations

Next, consider the link shears in two different Split-K EBF configurations: one a regular Split-K (Figure 3.4.7) and the other an alternating Split-K (Figure 3.4.8). The two configurations have been modeled and subjected to the same seismic lateral loading. Further, Figures 3.4.9 and 3.4.10 summarize axial forces in the various members for each configuration, while Figures 3.4.11 and 3.4.12 present lateral drifts for each. Several observations will provide insight into behavioral differences between a regular Split-K and an alternating Split-K eccentric braced frame.

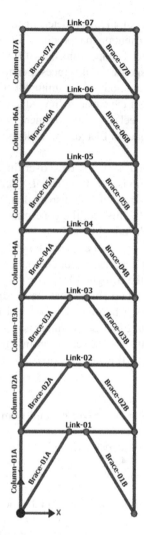

FIGURE 3.4.3   Sample Split-K eccentric braced frame.

1. The "Split-K" requires ductile links at seven levels.
2. The "alternating Split-K" requires ductile links at only four levels.
3. The shears in the "Split-K" at any given level are substantially less than the shears in the "alternating Split-K."
4. Brace forces are similar for the two systems.
5. Every other column axial load is substantially larger for the "alternating Split-K EBF."
6. Displacement demands are generally higher for the "alternating Split-K" EBF.
7. Axial forces in the beams outside the links are significantly higher for the regular "Split-K" EBF.

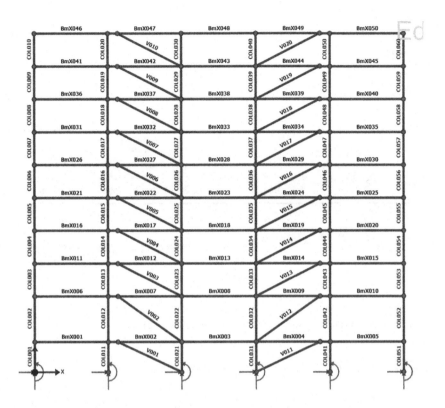

FIGURE 3.4.4 Sample alternate configuration eccentric braced frame.

In checking deflections for an eccentric braced frame (or any other lateral force resisting system), it may be necessary at times to separate the elastic from the plastic displacement components. One way to do this is to estimate the total and plastic component of displacement as given in Equations 3.4-8 and 3.4-9.

$$\delta_{tot} = C_d \delta_{xe} \tag{3.4-8}$$

$$\delta_p = (C_d - 1)\delta_{xe} \tag{3.4-9}$$

Example 3.4-3: EBF Link Design and Capacity-Limited Load Calculation

Figure 3.4.13 depicts an elevation view of an eccentric braced frame. The seismic loads, reduced by the force reduction factor, $R = 8$, are shown in the figure as well. The link shears at each level may be estimated using Equation 3.4-1.

$$(V_{Link})_{Roof} = \frac{68 \cdot 12.5}{30} = 28.3 \text{ kips}$$

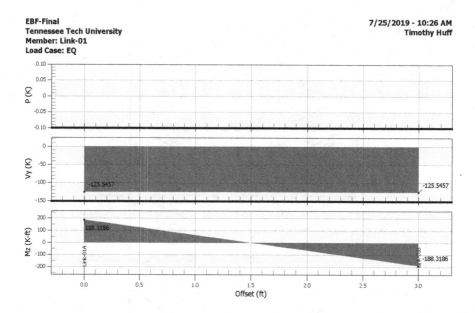

FIGURE 3.4.5   Axial load, shear, and moment in Split-K EBF link.

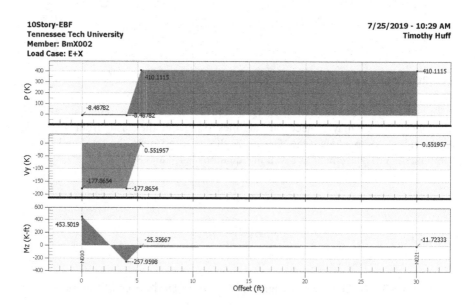

FIGURE 3.4.6   Axial load, shear, and moment in alternate configuration EBF link.

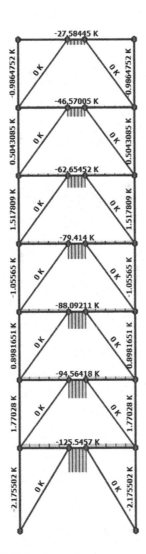

FIGURE 3.4.7 Split-K EBF link shears.

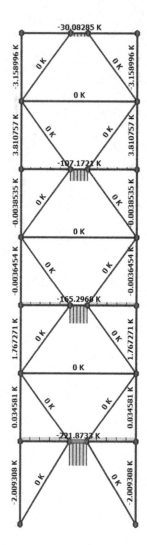

FIGURE 3.4.8   Alternating Split-K EBF link shears.

$$\left(V_{Link}\right)_{4th} = \frac{137 \cdot 12.5}{30} = 57.1\,\text{kips}$$

$$\left(V_{Link}\right)_{3rd} = \frac{182 \cdot 12.5}{30} = 75.8\,\text{kips}$$

$$\left(V_{Link}\right)_{2nd} = \frac{205 \cdot 14.0}{30} = 95.7\,\text{kips}$$

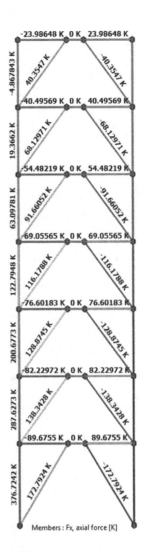

FIGURE 3.4.9   Split-K EBF axial forces.

For the link at the second floor, try a W14x53 made from A992 steel. From AISC 341-16, R_y = 1.1 for A992 steel. AISC 341-16 requires that webs of I-shaped links qualify as "highly ductile." Flanges of I-shaped links must qualify as "highly ductile," unless the link qualifies as a shear link, in which case it need only satisfy "moderately ductile" requirements. The web and flange requirements from Section D1.1 of AISC 341-16 are as follows for the proposed W14x53.

$$\lambda_{flange} = \frac{b}{t} = 6.11 < \left(\lambda_{hd}\right)_{flange} = 0.32\sqrt{\frac{29,000}{1.1 \cdot 50}} = 7.35$$

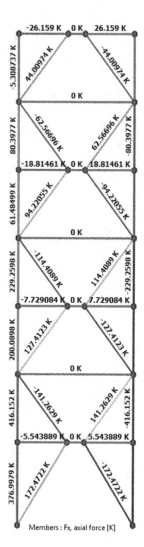

FIGURE 3.4.10 Alternating Split-K EBF axial forces.

$$\lambda_{web} = \frac{h}{t_w} = 30.9 < \left(\lambda_{hd}\right)_{web} = 2.57\sqrt{\frac{29,000}{1.1 \cdot 50}} = 59.0$$

Both the flange and the web of the A992 W14x53 qualify as highly ductile. Note that the above calculation for web ductility has an implicit assumption of an axial load approximately equal to zero in the link. Subsequent analysis by computer verified this assumption. In cases where the EBF configuration results in appreciable axial forces in the link, the ductility check must include this effect in accordance with AISC 341-16 Section D1.1.

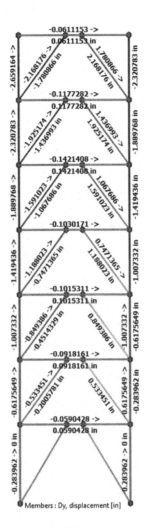

FIGURE 3.4.11 Lateral drifts for Split-K EBF.

Next, determine the link classification, the plastic moment, the design shear resistance, and the plastic rotation capacity. In checking the design shear resistance against the applied shear, a redundancy factor, $\rho = 1.3$, will be applied in accordance with ASCE 7-16 for the bottom story of this single-bay EBF.

$$M_P = 50 \cdot 87.1 = 4,355 \, \text{in} \cdot \text{kips}$$

$$A_{lw} = (13.9 - 2 \times 0.66)(0.370) = 4.65 \, \text{in}^2$$

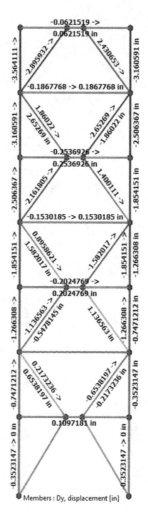

FIGURE 3.4.12   Lateral drifts for alternating Split-K EBF.

$$V_P = 0.6 \times 50 \times 4.65 = 139.6 \text{ kips}$$

$$e = 48 < 1.6 \cdot \frac{4,355}{139.6} = 49.9 \text{ ft} \rightarrow \text{Shear link} \rightarrow \gamma_P = 0.08 \text{ rad}$$

$$\frac{2 \times 4,355}{48} = 181.5 \text{ kips} > V_P = 139.6 \text{ kips}$$

$$\phi V_n = 0.90 \times 139.6 = 125.6 \text{ kips} > V_u = 1.3 \times 95.7 = 124.4 \text{ kips, OK}$$

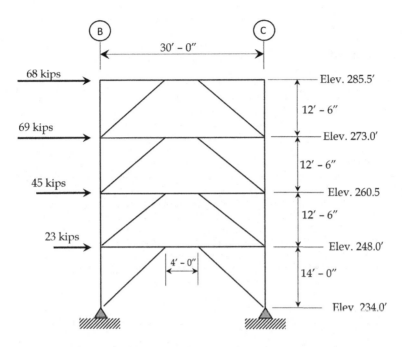

FIGURE 3.4.13 EBF link design and E_{cl} load calculation example.

Table 3.4.1 summarizes preliminary link designs for each level carried out with calculations similar to that for the second floor link. Note that the third floor link is slightly overstressed in the preliminary design. Nonetheless, the section is carried forward for further evaluation. Final design would require all links to be evaluated once all member sizes have been determined. Plastic drift capacity, Δ_P, in the table, is determined from Equation 3.4-2.

The capacity-limited loading condition, E_{cl}, may be determined by solving Equation 3.4-1 to determine the story shear at each level, which will produce link shears equal to $1.25 R_y V_n$ for the I-shaped links.

$$\left(V_{story}\right)_{Roof} = \frac{30}{12.5} \times 58.4 = 140.2 \text{ kips} \rightarrow \left(F_{Roof}\right)_{Ecl} = 140.2 \text{ kips}$$

$$\left(V_{story}\right)_{4th} = \frac{30}{12.5} \times 117.6 = 282.2 \text{ kips} \rightarrow \left(F_{4th}\right)_{Ecl} = 282.2 - 140.2 = 142.1 \text{ kips}$$

$$\left(V_{story}\right)_{3rd} = \frac{30}{12.5} \times 150.0 = 360.0 \text{ kips} \rightarrow \left(F_{3rd}\right)_{Ecl} = 360.0 - 282.2 = 77.8 \text{ kips}$$

$$\left(V_{story}\right)_{2nd} = \frac{30}{14.0} \times 192.0 = 411.4 \text{ kips} \rightarrow \left(F_{2nd}\right)_{Ecl} = 411.4 - 360.0 = 51.4 \text{ kips}$$

TABLE 3.4.1
Summary of EBF Link Designs

Level	Section	ϕV_n kips	V_u kips	Class	γ_P rad	Drift Δ_p in	$1.25R_yV_n$ kips
2nd	W14x53	125.6	124.4	Shear	0.08000	1.792	192.0
3rd	W8x58	98.2	98.5	Intermediate	0.07092	1.418	150.0
4th	W8x48	77.0	74.2	Intermediate	0.07542	1.508	117.6
Roof	W8x21	38.3	36.8	Flexural	0.02000	0.400	58.4

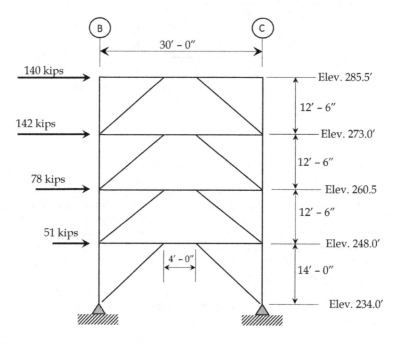

FIGURE 3.4.14 EBF capacity-limited loading.

Figure 3.4.14 shows the EBF with the capacity-limited loading. Comparison with Figure 3.4.13, the seismic load distribution for link design, should make it clear that elements other than the links will remain essentially elastic for a properly designed EBF during strong ground shaking.

Example 3.4-4: SCBF Brace Design and Capacity-Limited Load Calculation

Figure 3.4.15 is an elevation view of a special concentric braced frame (SCBF). Recall that, unlike the EBF, the SCBF fuse elements are the braces, yielding in

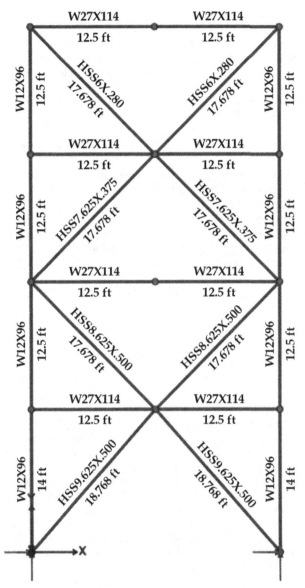

Brace Member Sizes Designed for:
1.2D + 0.5L + 0.2S + 1.3(E+X)

FIGURE 3.4.15   SCBF brace design and E_{cl} load calculation.

tension and either yielding or buckling in compression. The braces have been designed for the load combination that includes the seismic effect. The purpose of this example is to illustrate the determination of the capacity-limited load distribution for an SCBF. The design loading for this frame is higher than that for the EBF in the previous example because $R = 6$ for this SCBF, while $R = 8$ for the EBF. See Section 3.7 for detailed discussion on R-factors.

AISC 341-16 Section F2.3 requires that two cases be considered for the capacity-limited load in SCBF configurations:

(a) Brace forces correspond to expected strengths in compression and tension
(b) Brace forces correspond to expected strength in tension and post-buckling expected strength in compression

For the A500 HSS braces, $R_y = 1.4$ and $F_y = 42$ ksi. Expected strength in tension is given by Equation 3.4-10. Expected strength in compression is given by Equation 3.4-11. Post-buckling expected strength is to be taken as no more than 30% of the expected strength in compression.

$$P_{te} = R_y F_y A_g \tag{3.4-10}$$

$$P_{ce} = \text{Min} \begin{cases} R_y F_y A_g \\ F_{cre} A_g \\ \dfrac{}{0.877} \end{cases} \tag{3.4-11}$$

In Equation 3.4-11, F_{cre} is the critical compressive stress determined from AISC 360-16 Chapter E with $R_y F_y$ substituted for F_y. Effective brace lengths are required in the calculation of F_{cre}. For this example, KL is taken equal to $0.70L$. For final design, the engineer should calculate an accurate value for KL, taking into consideration the geometry of the connection and the fixity of the connection. Table 3.4.2 summarizes the calculations for the brace forces for the capacity-limited condition. Figures 3.4.16 and 3.4.17 depict the capacity-limited loading on the second floor beam for the two conditions to be considered.

Careful examination of Figures 3.4.16 and 3.4.17 reveal the importance of capacity-limited concepts for the design of the beams in SCBF configurations. The braces support the beams for dead and live load. The braces, in effect, load the beam during strong ground shaking.

TABLE 3.4.2
SCBF Brace Forces for Capacity-Limited Loading

Story	L_{brace} ft	A_g in²	r, in	KL in	F_e, ksi	$R_y F_y / F_e$	F_{cre} ksi	P_{te} kips	P_{ce} kips	P_{ce-pb} kips
1	18.768	13.4	3.24	157.7	120.9	0.4864	47.97	788	733	220
2	17.678	11.9	2.89	148.5	108.4	0.5424	46.86	700	636	191
3	17.678	7.98	2.58	148.5	86.4	0.6806	44.23	469	402	121
4	17.678	4.69	2.03	148.5	53.5	1.0993	37.12	276	198	60

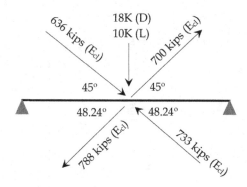

FIGURE 3.4.16   SCBF second floor beam capacity-limited load case 1.

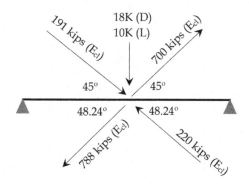

FIGURE 3.4.17   SCBF second floor beam capacity-limited load case 2.

3.5   CONNECTION DESIGN

While undergraduate steel design courses may introduce the student to simple bolted and welded connection design principles, a deeper knowledge of available connection types and theory will serve the beginning structural engineer well.

Of particular interest is the subject of using matching welds for connections. The weld metal strength required for "matched" welds should be based on the tensile strength, not the yield strength, of the base metal. So, for example, matched welds for A709 Grade 50W base metal (F_y = 50 ksi, F_u = 70 ksi, minimum) should be E70 electrodes.

The 15th edition of the AISC Steel Construction Manual includes useful tables that permit quick design for simple, shear-type connections.

Example 3.5-1: Simple (Shear) Connection Design

Given a W33x141 beam connected to the flange of a W14x120 column, both made from A992 steel, select an A36 double angle shear connection. The beam reactions are 55 kips due to dead load and 65 kips due to live load.

The factored (LRFD) beam reaction is:

$$R_u = 1.2 \times 55 + 1.6 \times 65 = 170 \text{ kips}$$

From page 10-28 of the AISC Construction Manual, select 3/8-inch thick angles with six rows of 1-inch diameter Group A bolts with threads not excluded from shear planes (A325-N) to achieve a design resistance of:

$$\phi R_n = 203 \text{ kips} > R_u = 170 \text{ kips, OK}$$

While the tables in the manual do incorporate design resistance values accounting for all possible failure modes of the angles and the bolts, the engineer must still check all possible failure modes of the connected beam web in accordance with Chapter J of AISC 360-16. It is critical that the engineer be well versed in the notes accompanying all design aid tables to ensure that all assumptions inherent in the tabulated values are valid for the particular case under consideration.

For non-seismic moment connections, a relatively simple connection incorporates beam flange plates welded to the column flanges with matched full penetration groove welds.

Example 3.5-2: Non-seismic Moment Connection

Suppose a W18x46 beam is to have end moment connection to a W12x87 column, both made from A992 steel. In addition to the column flange local bending, column web local yielding, column web crippling checks made below, web sidesway buckling, web compression buckling, and web panel zone shear should be checked, when applicable.

To permit advantageous welding position, the top flange plate of the beam should be at least 1 inch less in width than the beam flange. Similarly, the bottom flange plate of the beam should be at least 1 inch wider than the beam flange.

Design of the flange plates, in the example moment connection, will require consideration of tensile yield and rupture in accordance with Chapter D of AISC 360-16. Sample calculations follow. See Appendix C for detailed clarifications and calculations.

For the top plate, try a plate width of 4.5 inches since the beam flange width is 6.06 inches. Use A36 plates.

For gross section yielding:

$$A_{plate} \geq \frac{216 \text{ kips}}{0.90(36)} = 6.67 \text{ in}^2$$

$$t_{plate} \geq \frac{6.67}{4.50} = 1.48 \text{ inches} \rightarrow \text{try} 1.5 \text{ inches thick} \times 4.5 \text{ inches wide top plate}$$

For the bottom plate, try a plate width of 7.5 inches since the beam flange width is 6.06 inches. Use A36 plates.

For gross section yielding:

$$t_{plate} \geq \frac{6.67}{7.50} = 0.89 \text{ inches} \rightarrow \text{try} 1.0 \text{ inch thick} \times 7.5 \text{ inches wide bottom plate}$$

Try longitudinal welds only for the plate-to-beam flange connection. Case 4 from Table D3.1 in AISC 360-16 applies for the shear lag factor (U) calculation. Try weld lengths of 10 inches.

$$U_{top} = \frac{3 \times 10^2}{3(10)^2 + 4.5^2}\left(1 - \frac{1.5/2}{10}\right) = 0.866$$

$$\left(A_e\right)_{top} = 4.5 \times 1.5 \times 0.866 = 5.85 \text{ in}^2$$

$$\left(\phi R_n\right)_{rupture} = 0.75 \times 58 \text{ ksi} \times 5.85 \text{ in}^2 = 254 \text{ kips} > 216 \text{ kips, OK}$$

$$U_{bottom} = \frac{3 \times 10^2}{3(10)^2 + 7.5^2}\left(1 - \frac{1.0/2}{10}\right) = 0.800$$

$$\left(A_e\right)_{bottom} = 7.5 \times 1.0 \times 0.800 = 6.00 \text{ in}^2, \text{OK}$$

Appendix C contains hand calculations for stiffener requirements in accordance with AISC 360-16, Section J10.

Seismic connections for moment frames require special consideration in accordance with AISC 341-16 Section E1.6 for ordinary moment frames, E2.6 for intermediate moment frames, and E3.6 for special moment frames (American Institute of Steel Construction, 2016). In particular, it is advisable (and sometimes required) that prequalified connections described and detailed in AISC 358-16 (American Institute of Steel Construction, 2016) be incorporated into moment frame design for seismic loading.

Among the prequalified connections found in AISC 358-16 are (a) end plate connections, and (b) reduced beam section (RBS) connections.

Example 3.5-3: Reduced Beam Section Seismic Moment Connection

A W36x210 A992 beam is part of a special moment frame. Reduced beam section connection to W36x395 A992 columns is proposed. Design the RBS dimensions and ensure that the plastic hinge forms at the center of the RBS, as desired, rather than at the beam end.

The calculations for this example are found in Appendix D and should not be considered as complete, and the engineer is advised to consult AISC 341-16 and AISC 358-16 for a complete coverage of requirements.

A summary of the design parameters follows.

RBS Cut Dimensions: $a = 7$ inches $b = 24$ inches $c = 3$ inches
Beam-Column Connection Forces: $M_f = 3{,}376$ ft·k $V_u = 321$ kips

3.6 COMPUTER MODELING OF BUILDINGS

For this text, the educational version of IES VisualAnalysis has been used to demonstrate computer-modeling techniques for building structures.

Regardless of which software package is used, the engineer must be competent in each of the following.

Member Orientation: Proper orientation of the constructed members in the computer model is a necessary requirement in order to obtain reliable and relevant results. Figure 3.6.1 shows a simple frame with the left column and right column at orientations 90 degrees apart. The left column experiences major axis bending in the plane of the frame and the right column experiences minor axis bending.

Member Connectivity: Moment connections must be modeled as such and shear connections must have the appropriate moment releases. Figure 3.6.2 shows a frame with the beam moment-connected at the left column and shear-connected to the right column. Care must be taken in assigning member end moment releases. Should all

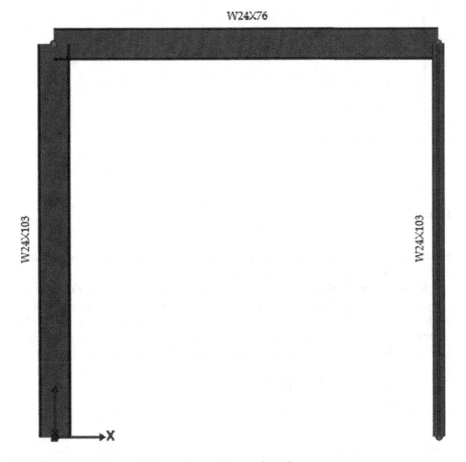

FIGURE 3.6.1 Frame with inconsistent column orientation.

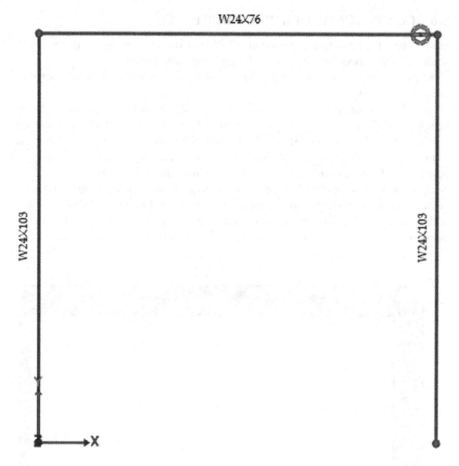

FIGURE 3.6.2   Moment versus shear connections.

members coming into a common joint have specified moment releases in the plane
of the members, instability will result in the analysis.

Element Type: Particularly for inelastic analysis, the engineer must know the
capabilities of the element employed. Many different formulations are available in
modern software. The element type used should be appropriate for the particular
member being modeled and a detailed study of all available types for a given computer program is warranted. These formulations may be broadly categorized as:

- Force-based, distributed plasticity
- Displacement-based, distributed plasticity
- Force-based, concentrated plasticity
- Displacement-based, concentrated plasticity
- Link-type elements

Link-type elements are often ideal for modeling seismic isolation bearings. Force-based elements frequently permit accurate modeling with a minimum of element subdivision. Displacement-based elements may be most appropriate for very short members. Distributed plasticity elements may be either fiber-based or plastic-hinge-based, the latter requiring the engineer to provide an estimate of the plastic hinge length.

3.7 VERTICAL SEISMIC LOAD DISTRIBUTION

ASCE 7-16 provides methods for computing seismic load on structures based on the "equivalent lateral force" (ELF) procedure. Buildings are assigned an importance factor, I_e, for seismic loading, based on the appropriate Risk Category. Table 3.7.1 summarizes the importance factor for both seismic loading and other extreme loading conditions.

Risk Category I structures are defined as those that pose negligible threat to human life in the event of failure and include barns, storage facilities, and similar structures. Risk Category III structures house large numbers of persons in one place and include theaters, lecture halls, elementary schools, prisons, and small health care facilities. Also included in Risk Category III are structures such as power-generating stations, water treatment plants, sewage treatment plants, and structures housing explosive, toxic, or otherwise hazardous materials. Risk Category IV includes hospitals, police stations, fire stations, and similar structures. For complete guidance on the assignment of appropriate Risk Categories, see ASCE 7-16 Section 1.5 and associated Commentary.

Analysis and design for seismic loading by the ELF method also requires the determination of the response modification factor (R), the overstrength factor (Ω_o), and the deflection amplification factor (C_d). Table 3.7.2 summarizes these factors from ASCE 7-16 for some commonly used LFRS configurations.

The response modification factor, R, reduces the design loading from the theoretical load that would preclude any damage or nonlinear behavior. Structures in modern codes, including ASCE 7-16, are typically designed to withstand controlled damage without collapse. The overstrength factor, Ω_o, is used as a means of ensuring that the planned ductile failure mode occurs and all other elements remain essentially elastic

TABLE 3.7.1
ASCE 7-16 Importance Factors

Risk Category	Snow I_s	Ice (Thick) I_i	Ice (Wind) I_w	Earthquake I_e
I	0.80	0.80	1.00	1.00
II	1.00	1.00	1.00	1.00
III	1.10	1.15	1.00	1.25
IV	1.20	1.25	1.00	1.50

TABLE 3.7.2

Seismic Design Factors in ASCE 7-16

LFRS	C_d	R	Ω_o
Steel Special Moment Frame (SMF)	5½	8	3
Steel Intermediate Moment Frame (IMF)	4	4½	3
Steel Ordinary Moment Frame (OMF)	3	3½	3
Concrete Special Moment Frame (SMF)	5½	8	3
Concrete Ordinary Moment Frame (OMF)	2½	3	3
Steel Eccentrically Braced Frame (EBF)	4	8	2
Steel Special Concentric Braced Frame (SCBF)	5	6	2
Steel Ordinary Concentric Braced Frame (OCBF)	3¼	3¼	2

during strong ground shaking. The deflection amplification factor, C_d, is used to estimate the effect of inelastic behavior on displacements of the structure.

Chapter 5 provides a detailed discussion on the development of seismic design response spectra control points, S_{DS}, S_{D1}, T_S, and T_L for a given site. Given these parameters, the ELF loading, in terms of total base shear V, is determined in accordance with Equations 3.7-1 and 3.7-2 and distributed vertically to the various floors in accordance with Equation 3.7-3. F_x is the ELF seismic force at level x, w_x is the "seismic weight" at level x, and h_x is the height above ground for level x. The "seismic weight" includes all permanent dead load and equipment loads in addition to at least 25% of the live load for storage areas. When the ground snow load is 30 psf or more, 20% of the design snow load is to be included in the "seismic weight" at the roof. Roof gardens should also be included when present. When a partition allowance is made in the design, the actual partition weight or a minimum of 10 psf is included in the "seismic weight."

The exponent k in Equation 3.7-4 is related to the natural period of the structure.

- $k = 1.00$ for $T = 0.50$ sec or less
- $k = 2.00$ for $T = 2.50$ sec or more
- Interpolate for T between 0.50 and 2.50 sec.

The natural period of the structure, T, also appears in these expressions. The natural period of the structure is unknown before design is complete, so ASCE 7-16 provides methods of estimating the natural period as T_a, presented here as Equation 3.7-5. Lower limits on C_s are also specified in ASCE 7-15. C_s shall not be taken lower than $0.044S_{DS}I_e$, nor shall C_s be taken lower than 0.01. Also, when S_1 (not S_{D1}) is 0.60 or greater, C_s shall not be taken less than $0.5S_1I_e/R$.

$$V = C_s \cdot W \qquad (3.7-1)$$

$$C_s = \frac{S_{DS}}{\left(\dfrac{R}{I_e}\right)} \leq \begin{cases} \dfrac{S_{D1}}{T\left(\dfrac{R}{I_e}\right)} & \text{for } T \leq T_L \\[4mm] \dfrac{S_{D1}T_L}{T^2\left(\dfrac{R}{I_e}\right)} & \text{for } T > T_L \end{cases} \qquad (3.7\text{-}2)$$

$$F_x = C_{vx}V \qquad (3.7\text{-}3)$$

$$C_{vx} = \frac{w_x h_x^k}{\displaystyle\sum_{i=1}^{N} w_i h_i^k} \qquad (3.7\text{-}4)$$

$$T_a = C_t h_n^x \qquad (3.7\text{-}5)$$

In Equation 3.7-5, for the approximate natural period of the structure, the coefficients depend on the type of construction. These are prescribed in ASCE 7-16 as follows:

- Steel eccentric braced frames: $C_t = 0.030$, $x = 0.75$
- Steel moment frames: $C_t = 0.028$, $x = 0.80$
- Steel buckling-restrained braced frames: $C_t = 0.030$, $x = 0.75$
- Concrete moment frames: $C_t = 0.016$, $x = 0.90$
- All other systems: $C_t = 0.020$, $x = 0.75$

Design story drift, Δ, is computed from the amplification factor, C_d, and the displacements, δ_{xe}, resulting from application of the ELF force distribution, F_x. Equation 3.7-6 defines the amplified design displacements, δ_x. Drift is computed as the difference in amplified displacement between the top and bottom of a given story.

$$\delta_x = \frac{C_d \delta_{xe}}{I_e} \qquad (3.7\text{-}6)$$

Drift limits prescribed in ASCE 7-16 are a function of the story height, h_{sx}, and are shown here in Table 3.7.3.

Load and resistance factor design (LRFD) load combinations from ASCE 7-16, which include seismic effects, are listed below.

- $U = 1.2D + E_v + E_h + L + 0.2S$
- $U = 0.9D - E_v + E_h$

The load factor on L in the first combination is permitted to equal 0.5, instead of 1.0, when the design live load is less than or equal to 100 psf, with the exception of garages or areas occupied as places of public assembly.

TABLE 3.7.3
ASCE 7-16 Drift Limits

	Risk Category		
Structure	I and II	III	IV
Structures, other than masonry shear wall structures, 4 stories or less above the base as defined in Section 11.2, with interior walls, partitions, ceilings, and exterior wall systems that have been designed to accommodate the story drifts	$0.025h_{sx}$	$0.020h_{sx}$	$0.015h_{sx}$
Masonry cantilever shear wall structures	$0.010h_{sx}$	$0.010h_{sx}$	$0.010h_{sx}$
Other masonry shear wall structures	$0.007h_{sx}$	$0.007h_{sx}$	$0.007h_{sx}$
All other structures	$0.020h_{sx}$	$0.015h_{sx}$	$0.010h_{sx}$

The vertical seismic load effect, E_v, and the horizontal seismic load effect, E_h, are defined in Equations 3.7-7 and 3.7-8. For the ELF method, Q_E is the effect due to the ELF load previously defined.

$$E_v = 0.20 \times S_{DS} \times D \qquad (3.7\text{-}7)$$

$$E_h = \rho Q_E \qquad (3.7\text{-}8)$$

The redundancy factor, ρ, depends on the structure configuration and details and on the seismic design category. The seismic design category is established in accordance with Tables 3.7.4 and 3.7.5.

In addition to the criteria in Tables 3.7.4 and 3.7.5, the following rules are to be applied:

- Risk Category I, II, or III structures with S_1 (not S_{D1}) ≥ 0.75 shall be assigned to Seismic Design Category E.
- Risk Category IV structures with S_1 (not S_{D1}) ≥ 0.75 shall be assigned to Seismic Design Category F.

TABLE 3.7.4
S_{DS}-Based Seismic Design Category (ASCE 7-16)

	Risk Category	
S_{DS} Range	I, II, and III	IV
$S_{DS} < 0.167$	A	A
$0.167 \le S_{DS} < 0.330$	B	C
$0.330 \le S_{DS} < 0.500$	C	D
$0.500 \le S_{DS}$	D	D

TABLE 3.7.5
S_{D1}-Based Seismic Design Category (ASCE 7-16)

S_{D1} Range	Risk Category	
	I, II, and III	IV
$S_{D1} < 0.067$	A	A
$0.067 \leq S_{D1} < 0.133$	B	C
$0.133 \leq S_{D1} < 0.200$	C	D
$0.200 \leq S_{D1}$	D	D

The redundancy factor, ρ, may be taken equal to 1.0 for structures in Seismic Design Categories B and C and for drift calculations and P-Delta evaluation for any seismic design category. For structures in Seismic Design Categories D, E, or F, ρ is 1.3 unless certain conditions are satisfied. Refer to section 12.3.4.2 of ASCE 7-16 for a detailed discussion of the redundancy factor for Seismic Design Categories D, E, and F.

The redundancy factor, ρ, for steel braced and steel moment frame structures in Seismic Design Categories D, E, and F may be taken equal to 1.00 if one (not both) of the following is satisfied:

1. In each story of the frame with more than 35% of the total base shear, the removal of a single brace (for braced frames) or the loss of moment resistance at both ends of a single beam (for moment frames) results in neither (a) no more than a 33% reduction in story strength, nor (b) an extreme torsional irregularity.
2. The structure is regular in plan, with at least two bays of perimeter LFRS framing on each side in each direction for all stories carrying more than 35% of the total base shear.

An extreme irregularity is defined as a story with stiffness less than 60% of that for the story above, or less than 70% of that for the average of the above three stories. Additional information and guidance on the assignment of appropriate redundancy factors requires careful attention to ASCE 7-16 requirements.

Modern seismic design philosophy relies on the identification of a ductile "fuse." The fuse is designed for the loadings and load combinations defined above to satisfy strength and drift criteria. All other elements are designed for a larger load, defined as either the overstrength load, $\Omega_o Q_E$, or the capacity-limited load, E_{cl}. The capacity-limited load is established based on an assumption that the fuse strength is higher than that used to determine the reliable resistance of the fuse.

The "fuses" in an eccentric braced frame are the links. The "fuses" in moment frames are plastic hinges in the beams (and possibly also in the columns at the ground floor). For a concentric braced frame, the "fuses" are the braces, yielding in tension and buckling in compression. For these systems, AISC 341-16 provides detailed design procedures for design of the "fuses" and for determination of the capacity-limited load, E_{cl}. Refer to Section 3.4 for examples.

Example 3.7-1: ELF Seismic Loading on a Four-Story Building

Figures 3.7.1, 3.7.2, and 3.7.3 are based on design examples presented in the AISC Seismic Design Manual. For this example, the structure is assumed an office building. Determine the "seismic weight," the ELF seismic base shear, the vertical distribution of the ELF seismic base shear, and the seismic design category. Data for this example are as follows:

- $S_S = 1.065$ g, $S_1 = 0.362$ g
- $S_{DS} = 0.852$ g, $S_{D1} = 0.702$ g, $T_L = 12$ sec, $T_S = 0.824$ sec, $T_O = 0.110$ sec
- $I_e = 1.00$
- Floor D = 85 psf
- Roof D = 68 psf
- Floor L = 50 psf
- Roof S = 20 psf
- Curtain Wall = 175 plf at each floor level and roof

For this example, the seismic weight will be taken as the dead load of the floors, the dead load of the roof, the dead load of the curtain wall, and a 10 psf partition allowance at each level.

$$w_2 = w_3 = w_4 = (0.085\text{ ksf} + 0.010\text{ ksf})$$

$$\times 9,000\text{ sf} + 0.175\text{ klf} \times 390\text{ ft} = 923\text{ kips}$$

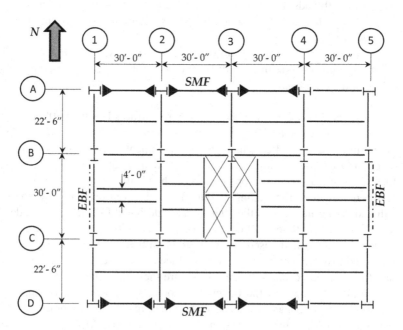

FIGURE 3.7.1 Typical floor plan for four-story building.

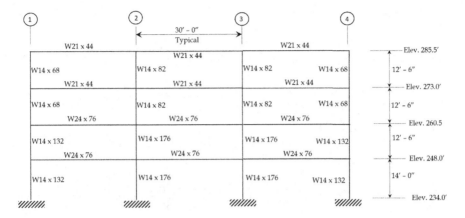

FIGURE 3.7.2 East-west special moment frame for four-story building.

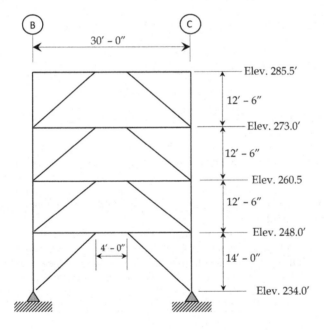

FIGURE 3.7.3 North-south eccentric braced frame for four-story building.

$$w_R = (0.068 \text{ ksf}) \times 9,000 \text{ sf} + 0.175 \text{ klf} \times 390 \text{ ft} = 680 \text{ kips}$$

$$\Sigma w = 3 \times 923 + 680 = 3,449 \text{ kips}$$

The roof height is 12.5 ft × 3 + 14 ft = 51.5 ft. This may now be used to determine the approximate natural period in each direction.

For the east-west SMF:

$$T_a = 0.030 \times 51.5^{0.75} = 0.58 \text{ sec}$$

$$k = 1.00 + (2.00 - 1.00) \times \frac{0.58 - 0.50}{2.50 - 0.50} = 1.04$$

For the north-south EBF:

$$T_a = 0.028 \times 51.5^{0.80} = 0.66 \text{ sec}$$

$$k = 1.00 + (2.00 - 1.00) \times \frac{0.66 - 0.50}{2.50 - 0.50} = 1.08$$

Rather than using different exponents, k, in the vertical distribution of forces, use the higher value of $k = 1.08$ since the periods are only very rough estimates at this point in the design process. The force reduction factor, R, is 8 for both the EBF and the SMF systems.

$$C_S = \frac{0.852}{\left(\frac{8}{1.00}\right)} = 0.1065$$

$$V = 0.1065 \times 3,449 = 367 \text{ kips}$$

Table 3.7.6 summarizes the calculations required to distribute the total base shear among the four levels of the building.

With $S_{DS} = 0.852 > 0.500$, the S_{DS}-based Seismic Design Category is D. With $S_{D1} = 0.702 > 0.200$, the S_{D1}-based Seismic Design Category is D. Therefore, the appropriate Seismic Design Category is D. When situations arise in which the S_{DS}-based and S_{D1}-based designations differ, the more severe case is to be used for the seismic design category. With $S_1 = 0.362 < 0.750$, Seismic Design Categories E and F do not apply.

Note here, as well, that the redundancy factors for the two framing systems will be different. With only a single-bay EBF on each north-south edge of the building, $\rho = 1.3$. With three bay moment frames on each east-west edge of the building, $\rho = 1.0$. See 12.3.4.2 of ASCE 7-16 for the rationale behind these values.

TABLE 3.7.6
Vertical Distribution of Horizontal Seismic Loading

Level	w, kips	h, ft	wh^k	C_{vx}	F_x
R	680	51.5	48,002	0.3333	122
4	923	39.0	48,256	0.3351	123
3	923	26.5	31,791	0.2208	81
2	923	14.0	15,960	0.1108	41
Sum	3,450		144,009	1.0000	367

3.8   HORIZONTAL SEISMIC LOAD DISTRIBUTION

Figure 3.8.1 depicts a plan view of a building floor, with the centers of mass and stiffness and the LFRS elements indicated. The analysis that follows assumes a rigid horizontal diaphragm capable of transmitting loads to the various LFRS elements in a rigid body manner. For structures with flexible horizontal diaphragms, the following analysis is not applicable.

Defining d_i as the perpendicular distance from the center of rigidity to LFRS element "i," the force distribution to the various LFRS elements may be computed as defined in Equations 3.8-1 through 3.8-4, taking careful note of the sign conventions used (e_X and e_Z are positive as shown).

$$V_1 = V_Z \times \frac{K_1}{K_1 + K_2} + \left(V_Z \times e_X - V_X \times e_Z\right) \times \frac{K_1 d_1}{\sum K_i d_i^2} \qquad (3.8\text{-}1)$$

$$V_2 = V_Z \times \frac{K_2}{K_1 + K_2} - \left(V_Z \times e_X - V_X \times e_Z\right) \times \frac{K_2 d_2}{\sum K_i d_i^2} \qquad (3.8\text{-}2)$$

$$V_3 = V_X \times \frac{K_3}{K_3 + K_4} - \left(V_Z \times e_X - V_X \times e_Z\right) \times \frac{K_3 d_3}{\sum K_i d_i^2} \qquad (3.8\text{-}3)$$

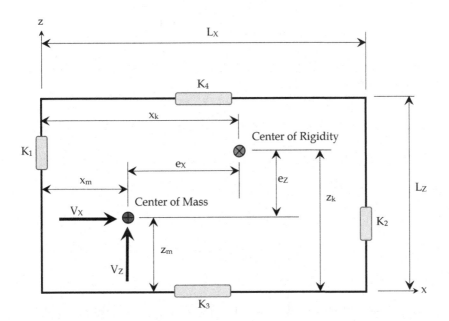

FIGURE 3.8.1   Floor plan with LFRS elements.

$$V_4 = V_X \times \frac{K_4}{K_3 + K_4} + \left(V_Z \times e_X - V_X \times e_Z\right) \times \frac{K_4 d_4}{\sum K_i d_i^2} \qquad (3.8\text{-}4)$$

The method is useful in evaluation of bidirectional effects from seismic loading. Different rules are used in combining orthogonal effects, V_X and V_Z. One method relies on the assumption that when one component is at its maximum, the orthogonal component is 30% of its maximum (the 100–30 rule). Another rule uses a 40% factor (the 100–40 rule) in place of 30%.

Example 3.8-1: Plan Torsional Effects on Load Distribution

In Figure 3.8.1, the following numerical values have been calculated and assigned. Determine the forces in each of the four LFRS elements. Use the 100–40 rule for combining orthogonal effects.

$K_1 = 40$ kips/in $K_2 = 80$ kips/in $K_3 = 50$ kips/in $K_4 = 100$ kips/in
$X_m = 54$ ft $Z_m = 33.75$ ft $V_X = 250$ kips $V_Z = 350$ kips
$L_X = 120$ ft $L_Z = 75$ ft

It is first necessary to determine the location of the center of rigidity.

$$X_k = \frac{40(0) + 80(120)}{40 + 80} = 80.0 \text{ ft}$$

$$Z_k = \frac{50(0) + 100(75)}{50 + 100} = 50.0 \text{ ft}$$

The perpendicular distance from each LFRS element to the center of rigidity, as well as the torsional rigidity factor (Σkd^2) and the eccentricities (e_X and e_Z), may now be determined:

$$d_1 = 80.0 \text{ ft}$$

$$d_2 = 120 - 80 = 40.0 \text{ ft}$$

$$d_3 = 50.0 \text{ ft}$$

$$d_4 = 75 - 50 = 25.0 \text{ ft}$$

$$\sum K_i d_i^2 = 40\left(80^2\right) + 80\left(40^2\right) + 50\left(50^2\right) + 100\left(25^2\right) = 571,500 \text{ kip}\cdot\text{ft}^2 \text{ /in}$$

$$e_x = 80.0 - 54.0 = 26.0 \text{ ft}$$

$$e_z = 50.0 - 33.75 = 16.25 \text{ ft}$$

Consider 100% of V_x (250 kips) in combination with 40% of V_z (140 kips).

$$V_1 = 140 \times \frac{40}{40+80} + \left(140 \times 26.0 - 250 \times 16.25\right) \times \frac{40 \times 80}{571,500} = 44.3 \text{ kips}$$

$$V_2 = 140 \times \frac{80}{40+80} - \left(140 \times 26.0 - 250 \times 16.25\right) \times \frac{80 \times 40}{571,500} = 95.7 \text{ kips}$$

$$V_3 = 250 \times \frac{50}{50+100} - \left(140 \times 26.0 - 250 \times 16.25\right) \times \frac{50 \times 50}{571,500} = 85.2 \text{ kips}$$

$$V_4 = 250 \times \frac{100}{50+100} + \left(140 \times 26.0 - 250 \times 16.25\right) \times \frac{100 \times 25}{571,500} = 164.8 \text{ kips}$$

Consider 40% of V_x (100 kips) in combination with 100% of V_z (350 kips).

$$V_1 = 350 \cdot \frac{40}{40+80} + \left(350 \cdot 26.0 - 100 \cdot 16.25\right) \cdot \frac{40 \times 80}{571,500} = 158.5 \text{ kips}$$

$$V_2 = 350 \cdot \frac{80}{40+80} - \left(350 \cdot 26.0 - 100 \cdot 16.25\right) \cdot \frac{80 \times 40}{571,500} = 191.5 \text{ kips}$$

$$V_3 = 100 \cdot \frac{50}{50+100} - \left(350 \cdot 26.0 - 100 \cdot 16.25\right) \cdot \frac{50 \times 50}{571,500} = 0.6 \text{ kips}$$

$$V_4 = 100 \cdot \frac{100}{50+100} + \left(350 \cdot 26.0 - 100 \cdot 16.25\right) \cdot \frac{100 \times 25}{571,500} = 99.4 \text{ kips}$$

Consider 100% of $-V_x$ (−250 kips) in combination with 40% of V_z (140 kips).

$$V_1 = 140 \cdot \frac{40}{40+80} + \left(140 \cdot 26.0 + 250 \cdot 16.25\right) \cdot \frac{40 \times 80}{571,500} = 89.8 \text{ kips}$$

$$V_2 = 140 \cdot \frac{80}{40+80} - \left(140 \cdot 26.0 + 250 \cdot 16.25\right) \cdot \frac{80 \times 40}{571,500} = 50.2 \text{ kips}$$

$$V_3 = -250 \cdot \frac{50}{50+100} - (140 \cdot 26.0 + 250 \cdot 16.25) \cdot \frac{50 \times 50}{571,500} = -117.0 \text{ kips}$$

$$V_4 = -250 \cdot \frac{100}{50+100} + (140 \cdot 26.0 + 250 \cdot 16.25) \cdot \frac{100 \times 25}{571,500} = -133.0 \text{ kips}$$

Consider 40% of $-V_x$ (−100 kips) in combination with 100% of V_z (350 kips).

$$V_1 = 350 \cdot \frac{40}{40+80} + (350 \cdot 26.0 + 100 \cdot 16.25) \cdot \frac{40 \times 80}{571,500} = 176.7 \text{ kips}$$

$$V_2 = 350 \cdot \frac{80}{40+80} - (350 \cdot 26.0 + 100 \cdot 16.25) \cdot \frac{80 \times 40}{571,500} = 173.3 \text{ kips}$$

$$V_3 = -100 \cdot \frac{50}{50+100} - (350 \cdot 26.0 + 100 \cdot 16.25) \cdot \frac{50 \times 50}{571,500} = -80.2 \text{ kips}$$

$$V_4 = -100 \cdot \frac{100}{50+100} + (350 \cdot 26.0 + 100 \cdot 16.25) \cdot \frac{100 \times 25}{571,500} = -19.8 \text{ kips}$$

Negative signs on the LFRS shears simply indicate forces in the opposite direction. Each LFRS should be designed for the maximum of the four cases considered.

$$V_1 = 176.7 \text{ kips}$$

$$V_2 = 191.5 \text{ kips}$$

$$V_3 = 117.0 \text{ kips}$$

$$V_4 = 164.8 \text{ kips}$$

3.9 MOMENT RESISTING COLUMN BASES

Undergraduate steel design courses often include basic design steps for base plates for columns subjected to axial compression only. The AISC Steel Construction Manual contains design assistance for such cases. When moment-resistant column bases are required, a design procedure based on that found in *AISC Design Guide 1: Base Plate and Anchor Rod Design* may be used and is the basis for the material presented here. Figure 3.9.1 shows a column and base plate subjected to simultaneous axial load and moment.

For solution of the problem, a uniform pressure distribution is assumed on the compression side. If the applied moment is small enough, it may not be necessary to develop tension in the anchor rods. By setting the uniform pressure equal to the design resistance, equilibrium considerations will determine whether tension in the anchor rods is required.

The design resistance based on bearing pressure on concrete is given here by Equation 3.9-1, taken from AISC 360-16 Section J8 for the case in which the base plate covers the full area of concrete support. For anchor rod design, Equation 3.9-2 provides the design resistance per anchor subjected to simultaneous tension and shear.

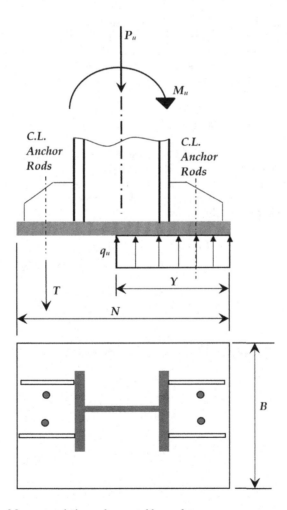

FIGURE 3.9.1 Moment resisting column and base plate.

A frequently used anchor rod material specification is ASTM F 1554 Grade 36 (F_u = 58 ksi), Grade 55 (F_u = 75 ksi), or Grade 105 (F_u = 125 ksi). From Table J3.2 of AISC 360-16, for F 1554 anchor rod with threads excluded from shear planes, F_{nt} = $0.75F_u$ and $F_{nv} = 0.563F_u$.

$$\phi P_p = 0.65\left[0.85 f_c' A_1\right]$$
(3.9-1)

$$F_{nt}' = 1.3F_{nt} - \frac{F_{nt}}{0.75F_{nv}} \cdot f_{rv} \le F_{nt}$$
(3.9-2)

Example 3.9-1: Moment Resisting Column Base

A W21x93 column has a moment resisting base required to carry loads of P_u and M_u. The proposed base plate dimensions and the loads are as follows. The base plate is conservatively assumed to be on the full area of concrete support. The concrete support has a concrete strength, f_c' = 6.5 ksi. The anchor centerline is 3 inches from the plate edge.

P_u = 610 kips M_u = 610 ft·kips V_u = 75 kips
B = 15 ft N = 30 ft

Appendix F is a set of hand calculations for the problem, excluding the effect of tension-shear interaction in the anchors. Check the tension-shear interaction to determine whether a revised anchor design is required.

For the 1.375-inch diameter F 1554 Grade 55 anchors, include the effect of the design shear and recheck the anchors.

$$F_{nt} = 0.75 \times 75 = 56.25 \text{ ksi}$$

$$F_{nv} = 0.563 \times 75 = 42.22 \text{ ksi}$$

$$A_b = \frac{\pi (1.375)^2}{4} = 1.485 \text{ in}^2 \text{ per anchor}$$

$$f_{rv} = \frac{75}{4 \times 1.485} = 12.63 \text{ ksi}$$

$$F_{nt}' = 1.3(56.25) - \frac{56.25}{0.75(42.22)} \cdot 12.63 = 50.7 \text{ ksi}$$

$$\phi R_n = 0.75 \times 50.7 \times 1.485 = 56.5 \text{ kips per anchor} \times 2 \text{ anchors}$$

$$= 113 \text{ kips} > 111 \text{ kips, OK}$$

Anchorage to concrete is covered in detail in ACI 318-19 (American Concrete Institute, 2019) Chapter 17. Anchors such as would be used for a column base include cast-in anchors, for which possible failure modes include:

- Steel failure in tension
- Steel failure in shear
- Anchor pullout
- Concrete breakout in tension
- Concrete breakout in shear
- Concrete blowout in tension
- Concrete pry-out in shear

The ACI check for a single steel anchor tension is given by Equation 3.9-3, where $A_{se,N}$ is the effective cross-sectional area of the anchor in tension and f_{uta} is the specified tensile strength of the anchor steel. f_{uta} is to be taken no larger than either $1.9f_{ya}$ or 125,000 psi.

$$\phi N_{sa} = \phi A_{se,N} \cdot f_{uta} \qquad (3.9\text{-}3)$$

The resistance factor, ϕ, for steel anchor tension is 0.75 for ductile anchors and 0.65 for brittle anchors in ACI 318-19. A ductile anchor is one for which the controlling failure mode is steel tension as opposed to any of the brittle concrete failure modes. Assuming a ductile failure mode, the ACI-based design tensile strength (in the absence of shear) for a single 1.375-inch diameter F 1554 Grade 55 anchor is calculated next. In particular, note that AISC uses the nominal bolt area, while ACI uses the effective tensile strength area tabulated in Table 7-17 of the AISC Steel Construction Manual.

$$\phi N_{sa} = \phi A_{se,N} \cdot f_{uta} = 0.75 \times 1.16 \times 75 = 65 \text{ kips}$$

For shear, the resistance factors are 0.65 for ductile anchors and 0.60 for brittle anchors. For cast-in anchors, the shear resistance is also based on effective area and tensile strength of the anchor. Therefore, the design shear strength (in the absence of tension) is easily determined:

$$\phi V_{sa} = \phi A_{se,V} \cdot f_{uta} = 0.65 \times 1.16 \times 75 = 56 \text{ kips}$$

For interaction between shear and tension, Equation 3.9-34 from ACI 318-19 is required.

$$\frac{N_u}{\phi N_n} + \frac{V_u}{\phi V_n} \leq 1.2 \qquad (3.9\text{-}4)$$

$$\frac{55.5}{65} + \frac{18.8}{56} = 1.19 < 1.2 \rightarrow OK$$

Therefore, whether AISC provisions or ACI provisions are used for anchor steel strength, the conclusion is similar in this example.

For basic pullout resistance, N_p, ACI provides equations for both headed bolts and for J or L bolts. Equation 3.9-5 is for headed bolts and Equation 3.9-6 is for J and L bolts. A_{brg} is the net bearing area of the headed bolt, d_a is the outside diameter of the hooked bolt, and e_h is the distance from the inner surface of the bolt shaft to the tip of the hook. The resistance factor for tensile pullout is 0.70. The "cracking factor," $\Psi_{c,P}$ is 1.4 when analysis indicates no service load cracking and 1.0 otherwise. The physical value specified for e_h must be no less than three times the anchor diameter, d_a. The value used for e_h in calculations must be no larger than 4.5 times the anchor diameter, d_a. It is easily observed that lengthening the hook beyond 4.5 anchor diameters provides no increase in pullout resistance. Headed anchors are generally superior to J and L bolts for pullout in tension.

$$\phi N_{pn} = \phi \Psi_{c,P} N_p = \phi \Psi_{c,P} \left[8 A_{brg} f'_c \right] \tag{3.9-5}$$

$$\phi N_{pn} = \phi \Psi_{c,P} N_p = \phi \Psi_{c,P} \left[0.9 f'_c e_h d_a \right] \tag{3.9-6}$$

The tension load on a single bolt for the current example is 55.5 kips. The required A_{brg} value for a headed anchor may be determined from this load. Assume cracked conditions for this example.

$$\phi N_{pn} = 0.70 \times 1.0 \times N_p = 0.70 \times 1.0 \left[8 \times A_{brg} \times 6.5 \right] \geq 55.5$$

$$\rightarrow A_{brg} \geq 1.525 \, in^2$$

The flat dimension of a standard square head bolt is 2.0625 inches, as tabulated in the AISC Manual (American Institute of Steel Construction, 2016) Table 7-18. The net bearing area is the area of the head minus the area of the bolt.

$$A_{brg} = 2.0625^2 - 1.485 = 2.769 \, in^2 > 1.525 \rightarrow OK$$

Had the area provided by a square head been insufficient, either a hex head, a heavy hex head, or a welded plate head could be specified to provide sufficient net bearing.

Note that the maximum design resistance achievable for a J or L bolt in this case would be insufficient, as demonstrated in the following calculation using the maximum permissible clear hook length, $e_h = 4.5 \times 1.375 = 6.19$ inches.

$$\phi N_{pn} = \phi \Psi_{c,P} N_p = 0.70 \cdot 1.0 \left[0.9 \cdot 6.5 \cdot 6.19 \cdot 1.375 \right]$$

$$= 34.9 \, kips < 55.5 \, kips \rightarrow No \, Good$$

To make a hooked bolt work in this case would require either an excessively large bolt diameter or an excessive concrete strength, or both.

For complete failure mode checks not included in this example, the reader is referred to ACI 318-19 Chapter 17.

3.10   CONCRETE MOMENT FRAMES

Both ACI 318-19 and ASCE 7-16 recognize concrete moment frames as viable systems for lateral force resistance. Concrete moment frames are classified as either ordinary, intermediate, or special moment frames. A subset of requirements from ACI 318-19 and ASCE 7-16 for each type of moment frame is summarized next. For a complete treatment of these systems, the reader is referred to Chapter 18 of ACI 318-19 and to Chapters 11, 12, and 14 of ASCE 7-16.

Ordinary concrete moment frames are permitted in only Seismic Design Category B, as defined in ASCE 7-16, Chapter 11. Intermediate concrete moment frames are permitted in Seismic Design Categories B and C. Special concrete moment frames are permitted in all Seismic Design Categories, A through F.

Beams in both ordinary and intermediate concrete moment frames must have at least two continuous bars at both top and bottom faces. Continuous bottom bars must have a total area at least equal to one-fourth of the maximum bottom bar area anywhere in the span. The continuous bars are to be anchored with development length based on the yield stress, f_y, in tension, at the face of all columns.

Intermediate concrete moment frames must also satisfy the requirements that (a) the positive moment strength at the column face must be no less than one-third of the negative moment strength at the column face, and (b) both negative and positive moment strength at any section along the span must be no less than one-fifth of the maximum moment strength provided at the column face.

Special concrete moment frames must additionally satisfy the requirements that (a) the positive moment strength at the column face must be no less than one-half of the negative moment strength at the face of the column and (b) both negative and positive moment strength at any section along the span must be no less than one-fourth of the maximum moment strength provided at the column face.

Beams in special concrete moment frames must have a clear span no less than four times the beam depth, d. Beam width in special concrete moment frames must be no less than either 10 inches or 30% of the beam height, h.

Design shear in beams for ordinary moment frames is to be taken as the lesser of (a) the shear associated with the development of nominal moment strength in reverse curvature at each beam end acting over the clear span, and (b) the maximum shear obtained from a structural analysis at the design load combinations with the seismic effect, E, amplified by the overstrength factor, Ω_o, which is equal to three for such systems.

Design shear in beams for intermediate moment frames is to be taken as the lesser of (a) the shear associated with the development of nominal moment strength in reverse curvature at each beam end acting over the clear span, and (b) the maximum shear obtained from a structural analysis at the design load combinations with the seismic effect, E, doubled.

Design shear in beams for special moment frames is to be taken as that shear associated with the development of probable, not nominal, moment strength in reverse curvature at each beam end acting over the clear span. Expected moment strengths are to be calculated using a bar stress equal to 1.25 times the yield stress and with a resistance factor, ϕ, equal to 1.0.

Columns in special moment frames must have the shortest cross-sectional dimension no less than 12 inches, and a ratio of short-to-long cross-sectional dimension of no less than 0.40. Column reinforcement must be no less than 1% and no greater than 6% of the cross-sectional area. Column strength at any joint in special moment frames must satisfy Equation 3.10-1, the so-called strong-column-weak-beam philosophy of seismic design in moment frame construction.

$$\sum M_{nc} \geq \frac{6}{5} \times \sum M_{nb} \tag{3.10-1}$$

M_{nc} is column nominal strength and M_{nb} is beam nominal strength. Both M_{nc} and M_{nb} are to be evaluated at the face of the beam-column joint.

Column design shear forces in special moment frames are to be determined using maximum probable flexural strengths, M_{pr}, at the top and bottom of the clear column height in reverse curvature bending. However, the design shear need not be taken greater than that obtained from analysis of the joint based on probable beam strengths, and certainly may not be taken less than that obtained from a structural analysis under the required load combinations.

Beam-column joint strength varies depending on the degree of confinement in ACI 318-19, Chapter 18. Equation 3.10-2 is a general expression for design joint strength using normal weight concrete, with the coefficient, x, varying between 8 for unconfined-noncontinuous joints and as much as 20 for fully confined and continuous joints. The concrete strength must be in units of psi and A_j is the effective joint area.

$$\phi V_n = 0.85\left(x \times \sqrt{f_c'} \times A_j\right) \tag{3.10-2}$$

Example 3.10-1 illustrates the design shear calculations for beams, columns, and joints in a special moment frame.

Example 3.10-1: Concrete Special Moment Frame

Structural analysis results for an interior column of a concrete special moment frame are shown in Figure 3.10.1. Actions for a typical beam are shown in Figure 3.10.2. The member actions shown for both the beam and the column are for the load combination, including seismic effects. The column shown is 48 inches square with 32 #9 longitudinal bars. Specified concrete strength is 7,500 psi and the reinforcement has a specified yield strength of 60 ksi. Beams framing into the column from all four sides are 24 inches wide and 42 inches deep at both the top and bottom of the column. Reinforcement patterns for the column and the

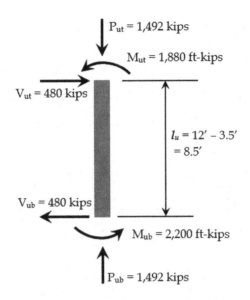

FIGURE 3.10.1   Member actions for SMF column.

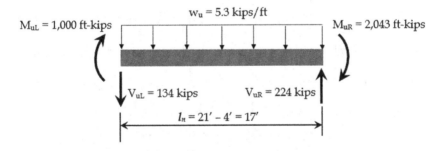

FIGURE 3.10.2   Member actions for SMF beam.

beam are presented in Figure 3.10.2. Floor-to-floor height is 12 ft and center-to-center column spacing is 21 ft. Axial load in the beams is found to be negligible (this will not always be the case). Preliminary reinforcement patterns for the beam and column are shown in Figure 3.10.3.

Note that the column reinforcement shown will require additional crossties to satisfy the requirement in ACI 318-19 that restrained transverse bars be spaced no farther than 14 inches apart in special moment frames.

A section analysis of the beam cross section using Response 2000 (Bentz, 2000) indicates a nominal flexural strength, M_n = 2,327 ft·kips. This initial section analysis of the beam was obtained using specified material strengths (7,500 psi concrete and 60 ksi steel) and a M/V ratio equal to 2,043/224 = 9.12 ft, as required by the member actions from the structural analysis. For the initial preliminary calculations, this flexural resistance is found to be adequate, since the cross section

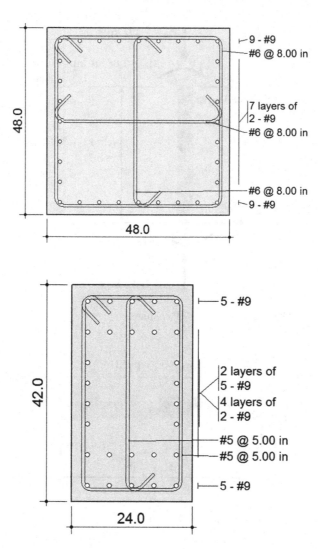

FIGURE 3.10.3 SMF beam and column preliminary reinforcement.

net tensile strain was found to be 0.009 > 0.005, indicating a tension-controlled beam section in accordance with ACI 318-19 Chapter 21 and a subsequent resistance factor of 0.90. Note that the beam reinforcement, for simplicity, has been arranged symmetrically in the top and bottom of the beam, though this will not generally be the case. In general, it will be necessary to perform the section analysis for both the left end (tension in the bottom reinforcement in this example) and the right end (tension in the top reinforcement in this example).

$$\phi M_n = 0.90 \times 2,327 = 2,094 \text{ ft} \cdot \text{k} > 2,043 \text{ ft} \cdot \text{k} \rightarrow \text{OK}$$

A second section analysis of the beam was performed to determine the probable flexural resistance, using an expected steel stress of 75 ksi and a resistance factor, ϕ = 1.0. The second sectional analysis indicates a probable resistance, M_{pr} = 2,815 ft·kips. With the same reinforcement at top and bottom faces of the beam, this probable resistance is then used to calculate the design shear based on the development of this resistance at both ends of the clear beam span. Setting both end moments equal to M_{pr} in Figure 3.10.2 and using statics to solve for the end shear produces the following:

$$V_{uR} = \frac{2,815 + 2,815 + 5.3(17)(17/2)}{17} = 376 \text{ kips}$$

For the column, an initial section analysis using specified material properties and an axial compression equal to the 1,492 kips shown in Figure 3.10.1 indicates a nominal flexural resistance, M_n = 3,590 ft·kips. The section analysis was performed at an M/V ratio equal to 2,200/480 = 4.58 ft to match the bottom conditions shown in Figure 3.10.1. The column section, with the large axial load, is compression-controlled with a net tensile strain, obtained from the sectional analysis, equal to the yield strain of 0.002. This requires a resistance factor equal to 0.65 in accordance with Chapter 21 of ACI 318-19. The column is found to satisfy strength requirements based on the member actions from the structural analysis shown in Figure 3.10.1.

$$\phi M_n = 0.65 \times 3,590 = 2,333 \text{ ft} \cdot \text{k} > 2,200 \text{ ft} \cdot \text{k} \rightarrow \text{OK}$$

A second section analysis of the column was performed using an expected steel stress of 75 ksi and a resistance factor, ϕ = 1.0, to determine the probable flexural strength, M_{pr} = 4,136 ft·kips. Again, using statics on the clear column height of 8.5 ft shown in Figure 3.10.1, and setting the end moments equal to M_{pr} yields the revised design shear for the column.

$$V_{ut} = V_{ub} = \frac{4,136 + 4,136}{8.5} = 973 \text{ kips}$$

These initial section analyses for the beam and column reveal the critical findings that (a) the beam must be designed for a shear of 376 kips (not the 224 kips indicated from the structural analysis for seismic effects), and (b) the column must be designed for a shear of 973 kips (not the 480 indicated by structural analysis). As iterations are completed, both shears will change.

For joint stress analysis, Figure 3.10.4 shows a typical joint with the top and bottom steel stressed to $1.25f_y$ = 75 ksi. The steel areas, A_{st} and A_{sb}, are both equal to 14.00 in² (14 #9 bars) in this simple example.

With continuity and confinement by beams on all four sides, the joint resistance may be determined from Equation 3.10-2 using a value for x equal to 20.

$$\phi V_n = \frac{0.85(20 \cdot \sqrt{7,500} \cdot 48^2)}{1000} = 3,392 \text{ kips} > 1,620 \text{ kips} \rightarrow \text{OK}$$

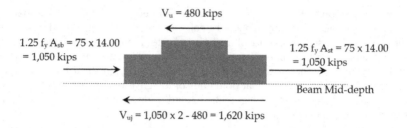

$V_u = 480$ kips

$1.25 f_y A_{sb} = 75 \times 14.00$
$= 1,050$ kips

$1.25 f_y A_{st} = 75 \times 14.00$
$= 1,050$ kips

Beam Mid-depth

$V_{uj} = 1,050 \times 2 - 480 = 1,620$ kips

FIGURE 3.10.4   SMF joint forces.

For nuclear facilities, ASCE 43-05 (ASCE, 2005) recognizes concrete special moment frames as acceptable systems, but explicitly prohibits both ordinary and intermediate moment frames. ASCE 43-05 refers the designer to ACI 349 (American Concrete Institute, 2013), not ACI 318, for member capacities.

3.11   CONCRETE SHEAR WALLS

Special and ordinary reinforced concrete shear walls are recognized as acceptable lateral force resisting systems in ASCE 7-16.

Special reinforced concrete shear walls of unlimited height are permitted in Seismic Design Categories B and C. Height limits for such systems are 160 ft in Seismic Design Categories D and E, and 100 ft in Seismic Design Category F.

Ordinary reinforced concrete shear walls of unlimited height are permitted in Seismic Design Categories B and C. Such systems are not permitted in Seismic Design Categories D, E, and F.

ACI 318-19, Chapter 11, outlines design requirements for ordinary walls. ACI 318-19, Chapter 18, Section 18.10, contains requirements for special reinforced concrete walls. Requirements for axial load and flexure are included in ACI. The focus here will be on the calculation of in-plane shear resistance for such systems.

For both ordinary walls not subjected to net axial tension and special walls, Equation 3.11-1 is specified for the design of in-plane shear resistance of walls constructed from normal weight concrete in ACI 318-19. The resistance factor for shear is 0.75 in ACI 318-19 Chapter 21. The units on concrete strength must be psi in the ACI equations.

$$\phi V_n = 0.75 \left(\alpha_c \sqrt{f_c'} + \rho_t f_{yt} \right) A_{cv} \leq 0.75 \left(8\sqrt{f_c'} \right) A_{cv} \qquad (3.11-1)$$

A few definitions are needed for Equation 3.11-1.

- $\alpha_c = 3$ for $h_w/l_w \leq 1.5$
- $\alpha_c = 2$ for $h_w/l_w \geq 2.0$
- Linear interpolation is to be used for intermediate values of h_w/l_w.
- l_w is the wall length in the direction of shear
- h_w is the clear height of wall

- ρ_t is the area of transverse reinforcement to gross concrete area perpendicular to that reinforcement
- ρ_l is the area of longitudinal reinforcement to gross concrete area perpendicular to that reinforcement
- f_{yt} is the specified yield strength of transverse reinforcement
- A_{cv} is the gross area of a concrete section bounded by the web thickness and the length of the section, with any openings deducted.

For multiple wall segments sharing the same lateral force, individual segments in special shear walls may have the upper limit on resistance calculated using a factor of 10 in place of the 8 on the right side of Equation 3.11-1.

Special walls often require boundary elements in accordance with ACI 318-19, Section 18.10.6.

4 Bridge Design

Limit states and load combinations for bridges are not the same as those adopted for buildings in modern codes. Resistance factors are different as well.

4.1 BRIDGE LOADS

Bridge loads include each of the following.

- CR – effects due to creep
- DC – dead load of components
- DW – dead load of wearing surface and utilities
- DD – downdrag
- EH – horizontal earth pressure
- EV – vertical earth pressure
- ES – earth surcharge load
- EL – miscellaneous locked-in forces
- PS – prestress forces
- SH – forces due to shrinkage
- BL – blast loading
- BR – vehicular braking force
- CE – vehicular centrifugal force
- CT – vehicular collision force
- CV – vessel collision force
- EQ – earthquake load
- FR – friction load
- IC – ice load
- IM – vehicular dynamic load allowance
- LL – vehicular live load
- LS – live load surcharge
- PL – pedestrian live load
- SE – force effect due to settlement
- TG – force effect due to temperature gradient
- TU – force effect due to uniform temperature
- WA – water load and stream pressure
- WL – wind on live load
- WS – wind load on structure

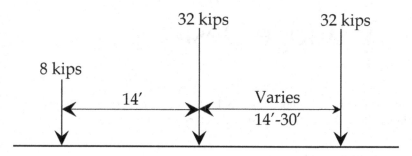

FIGURE 4.1.1   AASHTO design truck.

The design live loading for bridges in the AASHTO LRFD Bridge Design Specifications is a truck in combination with a uniform lane loading. The dynamic load allowance, *IM*, is taken as 15% of truck live load for the Fatigue Limit State and 33% of the truck live load for all other limit states. Figure 4.1.1 depicts the axle load and spacing for the design truck. The lane loading which acts simultaneously with the design truck is 0.640 kips per linear foot. The design truck plus lane is called the standard HL-93 loading. Impact is not applied to the uniform lane loading.

AASHTO also specifies a design tandem as two 25-kip axles spaced 4 ft apart. While the design tandem is lighter than the design truck, the tandem loading could certainly control design in situations where the design truck will not physically fit on the bridge, for reactions at interior piers in some cases of span arrangement, and for positive moment in spans for some span arrangements. The transverse wheel spacing is 6 feet for both the truck and the tandem loadings.

For interior reactions and negative moments at intermediate piers or bents, 90% of the effect from two trucks plus the lane loading are prescribed in AASHTO.

For the Fatigue Limit State, defined as the fatigue truck only (with no simultaneous lane loading), the spacing between the two 32-kip axles is constant and equal to 30 ft.

The standard lane width in AASHTO is 12 ft. Traffic is placed in design lanes with wheels no closer than 2 ft to the edge of a design lane. An exception to this is the design of the roadway deck, for which the wheel is placed 1 ft from the face of the barrier. The number of design lanes is taken as the integer portion of the roadway width divided by 12. Design for partial lanes is not required.

4.2   LIMIT STATES AND LOAD COMBINATIONS

AASHTO (AASHTO, 2017) limit states and load factors are summarized in Tables 4.2.1 through 4.2.4.

The purpose here is not to give a comprehensive treatment of the development of loading on bridge structures, but to provide a discussion and example calculations for those items that can often present unique difficulties. Subsections to follow include

TABLE 4.2.1
AASHTO Strength Limit States

Limit State	DC DD DW EH EV ES EL PS CR SH	LL IM CE BR PL LS	WA	WS	WL	FR	* TU	TG	SE
Strength I	γ_p	1.75	1.00	–	–	1.00	0.50/1.20	γ_{TG}	γ_{SE}
Strength II	γ_p	1.35	1.00	–	–	1.00	0.50/1.20	γ_{TG}	γ_{SE}
Strength III	γ_p	–	1.00	1.00	–	1.00	0.50/1.20	γ_{TG}	γ_{SE}
Strength IV	γ_p	–	1.00	–	–	1.00	0.50/1.20	–	–
Strength V	γ_p	1.35	1.00	1.00	1.00	1.00	0.50/1.20	γ_{TG}	γ_{SE}

Strength I – Basic load combination relating to the normal vehicular use of the bridge without wind.

Strength II – Load combination relating to the use of the bridge by owner-specified special design vehicles.

Strength III – Load combination relating to the bridge exposed to the design wind speed.

Strength IV – Load combination emphasizing dead load force effects in bridge superstructures.

Strength V – Load combination relating to normal vehicular use of the bridge with wind of 80 mph velocity.

*: The larger factor shall be used for deformations and the smaller factor for forces.

treatment of (a) *TU* (uniform temperature change), (b) *BR* (braking force), (c) *CE* (centrifugal force), (d) *WS* (wind on structure), and (e) *EQ* (seismic loading). Live load effects are treated separately in Section 4.3.

4.2.1   TU LOADING ON BRIDGES

Thermal effects on bridge structures can be problematic and difficult to predict. AASHTO includes two methods for computing thermal expansion and contraction requirements for bridges. In method A, the design temperature range is taken to be as follows:

- –30°F to 120°F, steel structures in cold climates
- 0°F to 120°F, steel structures in moderate climates
- 0°F to 80°F, concrete structures in cold climates
- 10°F to 80°F, concrete structures in moderate climates

TABLE 4.2.2
AASHTO Service Limit States

Limit State	DC DD DW EH EV ES EL PS CR SH	LL IM CE BR PL LS	WA	WS	WL	FR	TU*	TG	SE
Service I	1.00	1.00	1.00	1.00	1.00	1.00	1.00/1.20	γ_{TG}	γ_{SE}
Service II	1.00	1.30	1.00	–	–	1.00	1.00/1.20	–	–
Service III	1.00	γ_{LL}	1.00	–	–	1.00	1.00/1.20	γ_{TG}	γ_{SE}
Service IV	1.00	–	1.00	1.00	–	1.00	1.00/1.20	–	1.00

Service I – Load combination relating to the normal operational use of the bridge with a 70-mph wind.

Service II – Load combination intended to control yielding of steel structures and slip of slip-critical connections.

Service III – Load combination for longitudinal analysis relating to tension in prestressed concrete superstructures.

Service IV – Load combination relating only to tension in prestressed concrete columns for crack control.

*: The larger factor shall be used for deformations and the smaller factor for forces.

TABLE 4.2.3

AASHTO Extreme Event and Fatigue Limit States

Limit State	DC DD DW EH EV ES EL PS CR SH	LL IM CE BR PL LS	WA	WL	FR	EQ	BL*	IC*	CT*	CV*
Extreme Event I	1.00	γ_{EQ}	1.00	–	1.00	1.00	–	–	–	–
Extreme Event II	1.00	0.50	1.00	–.	1.00	–	1.00	1.00	1.00	1.00
Fatigue I LL, IM, CE	–	1.75	–	–	–	–	–	–	–	–
Fatigue II LL, IM, CE	–	0.80	–	–	–	–	–	–	–	–

Fatigue I – Fatigue and fracture load combination related to infinite load-induced fatigue life.

Fatigue II – Fatigue and fracture load combination related to finite load-induced fatigue life.

* Use only one at a time.

TABLE 4.2.4
AASHTO Variable Load Factors

	Load Factor, γ_P	
	Maximum	Minimum
DC – Strength I, II, III, and V	1.25	0.90
DC – Strength IV	1.50	0.90
DW	1.50	0.65
EH – Active	1.50	0.90
EH – At rest	1.35	0.90
EV – Overall stability	1.00	NA
EV – Retaining walls and abutments	1.35	1.00
ES	1.50	0.75

	Load factor, γ_{LL}, for Service III Limit State
PPC with refined losses/elastic gain	1.00
All other PPC	0.80

Values for the coefficient of thermal expansion, α_T, are typically taken to be 0.0000060/°F for concrete and 0.0000065/°F for steel.

Example 4.2-1: Thermal Expansion and Contraction Requirements

Figure 4.2.1 shows the elevation of a bridge constructed on State Route 26 over Sligo Road and Center Hill Lake in Dekalb County, TN. Figure 4.2.2 is a photograph taken during construction of Abutment No. 1 at the beginning of the bridge.

To resist large vertical loads in challenging subsurface conditions as well as lateral earth pressures and longitudinal forces from other sources, Abutment 1 (approximately 25 ft tall) was designed to be supported on multiple rows of piles with 51 total piles. This was judged to result in a substructure at Abutment 1 of such large stiffness that Abutment 1 was considered the center of stiffness, with thermal expansion and contraction for the entire bridge taken at Abutment 2. The thermal expansion and contraction movement requirements for the bridge bearings at Abutment 2 follow. Dekalb County, TN, is considered a moderate climate in AASHTO maps for *TU* loading. The bridge girders are steel welded plate girders. Note also that the deformation-based load factor for *TU* loading is 1.20. End span lengths are each 270-ft and interior spans are each 335-ft.

$$(\Delta_{TU})_{tot} = 1.20[0.0000065 \times 120 \times 1,535 \times 12] = 17.2 \text{ inches}$$

FIGURE 4.2.1   Elevation of State Route 26 over Sligo Road and Center Hill Lake.

FIGURE 4.2.2   Sligo Bridge Abutment No. 1 during construction.

An expansion joint and expansion bearings were provided at Abutment 2 with a total movement capacity of 18 inches. Figure 4.2.3 depicts an excerpt from the bridge plans for the expansion bearing at Abutment 2.

The total reaction at Abutment 2, per bearing, is 498 kips for the Service Limit State and 660 kips for the Strength Limit State. Device dimensions are as follows:

- BL = 36 inches
- BW = 29 inches
- CL = 28 inches
- CW = 27 inches
- SL = 28 inches
- SW = 47 inches

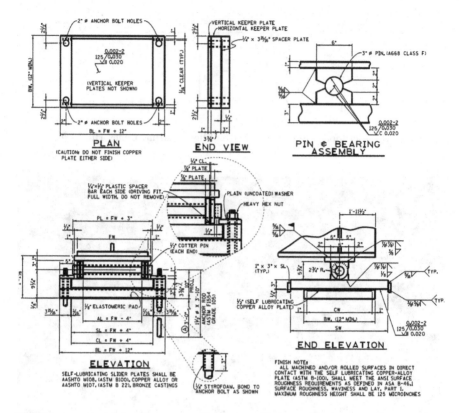

FIGURE 4.2.3   Abutment 2 expansion bearings for the Sligo Bridge.

- PL = 27 inches
- AL = 20 inches

The total movement capacity of the device is $SW - BW = 47 - 29 = 18$ inches, as required.

The coefficient of friction for the sliding surface is specified in AASHTO to be 0.10 for bronze and 0.40 for all other self-lubricating surfaces. A value of 0.40 was used in the design of the bearing.

AASHTO limits the stress on the sliding surface to 2 ksi at the Service Limit State. AASHTO criteria for bearing on concrete and anchor rod shear are provided for the Strength Limit State. Appendix E contains hand calculations for the bearing pressures on the sliding self-lubricating plate and the concrete, as well as anchor rod calculations.

4.2.2   BR Forces

Braking forces (BR) are longitudinal forces (in the direction of traffic) accounting for effects imparted by braking vehicles on the bridge. Braking forces in the AASHTO

LRFD Bridge Design Specifications are applied 6 ft above the riding surface in all lanes carrying traffic in the same direction and are defined as the larger of:

- 25% of design truck
- 25% of design tandem
- 5% of (design truck plus design lane)
- 5% of (design tandem plus design lane)

Example 4.2-2: Braking Force Calculation

Figure 4.2.4 is one-half of a dual structure. Each bridge consists of three spans, 195–230–195 ft, and carries one-directional traffic. The braking forces are to be carried by the two bents, with contribution from the abutments ignored. Bent 1 columns are 33 ft tall and Bent 2 columns are 42 ft tall. Estimate the required braking force at each bent. The assumption of zero contribution from the abutments would be appropriate with expansion joints and bearings at both abutments. In integral abutment bridges, the abutments may be so stiff as to carry a large portion of the braking forces.

Determine the number of traffic lanes.

$$N_L = Int\left(\frac{52.5 - 2 \times 7.5 \div 12}{12}\right) = 4\,\text{lanes}$$

Since the force at an intermediate bent is under consideration, consider the loading as two trucks plus the design lane uniform loading. Determine the braking force for each of the prescribed cases:

$$BR_A = 0.25[72\,k \times 2\,\text{trucks}](0.90) = 32.4\,k\,\text{per lane}$$

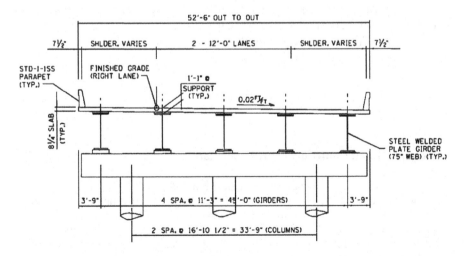

FIGURE 4.2.4 Braking force calculation example.

$$BR_B = 0.05\left[72\,\text{k} \times 2\,\text{trucks} + 0.64 \times 620\right](0.90) = 24.3\,\text{k per lane}$$

BR_A controls: With four lanes designated for traffic all in the same direction and applying the multi-presence factor, $m = 0.65$, for four loaded lanes, the braking force may be determined. Since the bents are at different heights, the braking force will not be equally distributed between them. Given that the stiffness of a bent is inversely proportional to the cube of the column height, determine the braking force at each bent. Take KR_1, the relative stiffness of Bent 1, equal to 1.0.

$$KR_2 = 1.0\left(\frac{33}{42}\right)^3 = 0.485$$

$$BR_{TOT} = 4\,\text{lanes} \times 0.65 \times 32.4\,\text{kips per lane} = 84.2\,\text{kips}$$

The braking forces at Bent 1, BR_1, and Bent 2, BR_2, are thus easily determined:

$$BR_1 = 84.2\left(\frac{1.0}{1.0 + 0.485}\right) = 56.7\,\text{kips}$$

$$BR_2 = 84.2\left(\frac{0.485}{1.0 + 0.485}\right) = 27.5\,\text{kips}$$

4.2.3   CE FORCES

Centrifugal forces (CE) arise from moving traffic on curved structures. AASHTO defines the centrifugal force as a factor, C, multiplied by the weight, presumably a 72-kip truck. Equation 4.2-1 gives the factor C. The design velocity, v, must be in units of ft/sec. The curve radius, R, must be in units of ft. The acceleration of gravity, g, is 32.2 ft/s². The factor, f, is 1.0 for the Fatigue Limit State and 4/3 for all other limit states:

$$C = f\frac{v^2}{gR} \tag{4.2-1}$$

Example 4.2-3: Centrifugal Force Calculation

Figure 4.2.5 is a cross section of the flyover ramp depicted in Figure 4.2.6. The flyover is located west of Nashville, TN at the Interstate 40/White Bridge Road interchange. The curve radius measured to the Ramp baseline is 755 ft. The design speed for the ramp is 50 mph. The center of the curve is to the left of the cross section in Figure 4.2.5. Determine the Centrifugal Forces, F_{CE}, at the Fatigue Limit State and at the Strength Limit State.

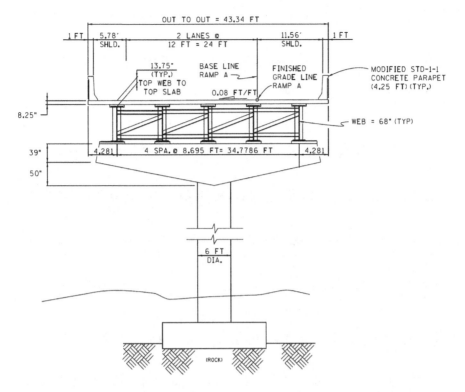

FIGURE 4.2.5   Flyover ramp cross section.

FIGURE 4.2.6   Flyover ramp at White Bridge Road Interchange (Google Earth).

Determine the number of design lanes:

$$N_L = \text{Integer}\left(\frac{43.34 - 2}{12}\right) = 3 \text{ lanes}$$

Determine the radius of curvature for the innermost lane, Lane 1.

$$R_1 = 755 - 24 - 5.78 + 6 = 731.2 \text{ ft}$$

The radius of curvature for each successive lane will be 12 ft in excess of that for the previous lane:

$$R_2 = 731.2 + 12 = 743.2 \text{ ft}$$

$$R_3 = 743.2 + 12 = 755.2 \text{ ft}$$

The design velocity, converted to ft/sec, is:

$$v = 50 \text{ mph} \times \frac{5,280 \text{ ft}}{\text{mile}} \times \frac{\text{hr}}{3,600 \text{ sec}} = 73.3 \text{ fps}$$

The factor, C, for each of the three lanes may now be determined:

$$C_1 = 1.0 \frac{73.3^2}{32.2 \times 731.2} = 0.228, \text{ Fatigue Limit State}$$

$$C_1 = \frac{4}{3} \cdot \frac{73.3^2}{32.2 \times 731.2} = 0.305, \text{ All other limit states}$$

$$C_2 = 1.0 \frac{73.3^2}{32.2 \times 743.2} = 0.225, \text{ Fatigue Limit State}$$

$$C_2 = \frac{4}{3} \cdot \frac{73.3^2}{32.2 \times 743.2} = 0.300, \text{ All other limit states}$$

$$C_3 = 1.0 \frac{73.3^2}{32.2 \times 755.2} = 0.221, \text{ Fatigue Limit State}$$

$$C_3 = \frac{4}{3} \cdot \frac{73.3^2}{32.2 \times 755.2} = 0.295, \text{ All other limit states}$$

For the Fatigue Limit State, defined as a single-loaded lane with a single truck:

$$\left(F_{CE}\right)_{Fatigue} = 0.228 \times 72 \text{ kips} = 16.4 \text{ kips}$$

For the Strength Limit State, with three lanes loaded and the multi-presence factor, m, taken equal to 0.85:

$$\left(F_{CE}\right)_{Strength} = 0.85 \times \left(0.305 + 0.300 + 0.295\right) \times 72 \text{ kips} = 55.1 \text{ kips}$$

4.2.4   WS Forces

Wind on structure forces for bridges in AASHTO depend on 3-sec gust wind speeds. This is a departure from some past editions of the AASHTO Specifications, which were based on fastest-mile wind speeds. Hence, it is stressed that mixing codes and specifications should always be avoided.

Design wind speed varies depending on the limit state being considered. This is also a departure from past editions of the AASHTO Specifications.

- For the Strength III Limit State, the design wind speed, V, is taken from maps of 3-sec gust wind speed having a 7% probability of exceedance in 50 years.
- For the Strength V Limit State, $V = 80$ mph.
- For the Service I Limit State, $V = 70$ mph.
- For the Service IV Limit State, V is taken as 75% of that used for the Strength III Limit State.

Multiple angles of wind load direction are considered in AASHTO provisions. With wind perpendicular to the bridge, a vertical uplift applied to the deck, in addition to horizontal pressure on the bridge, is included. The vertical pressure is 20 psf applied at the ¼ point of the deck for the Strength III Limit State and 10 psf applied at the ¼ point of the deck for the Service IV Limit State.

The wind speed, V, produces a pressure, P_Z, given by Equation 4.2-2. The gust factor, G, is 1.00 for bridges.

$$P_Z = 2.56 \times 10^{-6} \times V^2 K_Z G C_D \qquad (4.2\text{-}2)$$

- P_Z = wind pressure, ksf
- V = 3-sec gust, mph
- K_Z = exposure and elevation coefficient (use equations for Strength III and Service IV, use 1.0 for other limit states)

- Z is the height above surrounding ground or water and is never to be taken less than 33 ft.

The exposure and elevation coefficient depends on the exposure condition and the relationships are presented here in Equations 4.2-3 through 4.2-5:

$$K_Z(B) = \frac{\left[2.5 \ln\left(\dfrac{Z}{0.9834}\right) + 6.87\right]^2}{345.6} \qquad (4.2\text{-}3)$$

$$K_Z(C) = \frac{\left[2.5 \ln\left(\dfrac{Z}{0.0984}\right) + 7.35\right]^2}{478.4} \qquad (4.2\text{-}4)$$

$$K_Z(D) = \frac{\left[2.5 \ln\left(\dfrac{Z}{0.0164}\right) + 7.65\right]^2}{616.1} \qquad (4.2\text{-}5)$$

Exposure conditions are similar to those found in ASCE 7-16 for buildings. Exposure B may be generally described as areas with closely spaced obstructions, Exposure C as open terrain with widely scattered obstructions, and Exposure D as flat unobstructed areas and water surfaces. Refer to the AASHTO Bridge Design Specifications for more detailed information on Exposure categories.

The drag coefficient, C_D, is 1.3 for the windward side of a bridge superstructure, and 1.6 for bridge substructures. The entire wind pressure is applied to the windward side of both the superstructure and the substructures in the current AASHTO Bridge Design Specifications.

Example 4.2-4: Wind Load on Structure for Sligo Bridge

Figure 4.2.7 is an elevation of State Route 26 over Sligo Road and Center Hill Lake in Dekalb County, TN. Figure 4.2.8 is a cross section of the bridge. Determine the Strength III Limit State WS loading on Pier Number 2 of the bridge with the water

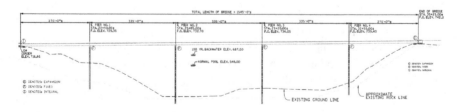

FIGURE 4.2.7 Elevation of the Sligo Bridge.

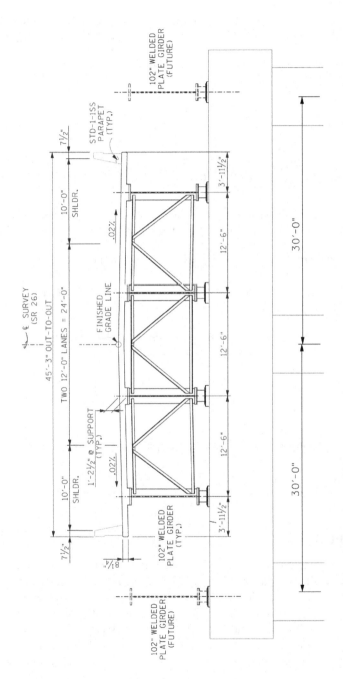

FIGURE 4.2.8 Cross section of the Sligo Bridge.

surface at normal pool elevation. End spans are each 270-ft and interior spans are each 335-ft.

The column diameter is 8 ft. The total superstructure depth is 12.8 ft from bottom of girder to top of parapet. From the wind maps for 3-sec gust in AASHTO, the design wind speed is $V = 115$ mph. With a large lake surface, use Exposure D. Determine the vertical load based on the future deck width of 45.25 ft + 12.5 ft × 2 = 70.25 ft.

$$Z = 742 - 648 = 94 \text{ ft}$$

$$K_Z(D) = \frac{\left[2.5\ln\left(\dfrac{94}{0.0164}\right) + 7.65\right]^2}{616.1} = 1.392$$

For load on the superstructure, assume that Pier 2 carries one-half of each adjacent span for wind load distribution from the superstructure. For horizontal and vertical loads on the Pier from the superstructure, use $(F_{WS})_H$ and $(F_{WS})_V$ respectively. For uniform load on the windward column of the substructure, use $(w_{WS})_H$:

$$P_Z = 2.56 \times 10^{-6} \times 115^2 (1.392)(1.00)(1.3) = 0.061\,\text{ksf} = 62 \text{ psf}$$

$$(F_{WS})_H = 0.062 \text{ ksf} \times 12.8' \times \left(\frac{1}{2} \times 335 + \frac{1}{2} \times 335\right) = 266 \text{ kips}$$

$$(F_{WS})_V = 0.020 \text{ ksf} \times 70.25' \times \left(\frac{1}{2} \times 335 + \frac{1}{2} \times 335\right) = 471 \text{ kips}$$

For the substructures:

$$P_Z = 2.56 \times 10^{-6} \times 115^2 (1.392)(1.00)(1.6) = 0.075 \text{ ksf} = 75 \text{ psf}$$

$$(w_{WS})_H = 0.075 \text{ ksf} \times 8 \text{ ft} = 0.600 \text{ klf}$$

4.2.5 EQ EFFECTS

Current seismic loading in AASHTO for bridges is based on ground shaking having a 7% probability of exceedance in 75 years. This design basis has been under regular scrutiny in the past two decades, and the bridge engineer will be well served in keeping abreast of design requirements as they change. Current seismic design criteria for bridges are based on uniform hazard, rather than risk-targeted, ground motion.

Building design in ASCE 7-16 is based on two-thirds of the ground shaking having a 2% probability of exceedance in 50 years, and the design ground motion is risk-targeted.

Options available to the structural engineer for the seismic design of bridges include:

(a) Force-based design in accordance with the AASHTO LRFD Bridge Design Specifications
(b) Displacement-based design in accordance with the AASHTO Guide Specification for LRFD Seismic Bridge Design
(c) The AASHTO Guide Specification for Seismic Isolation Design

For conventional design, in which the inelastic behavior is accommodated in either the plastic hinging of substructure columns or plastic behavior of ductile end super-structure diaphragms, the displacement-based design is the more advanced and appropriate engineering solution.

The design response spectrum is defined by the control points, A_S, S_{DS}, S_{D1}, T_S, and T_o. See Chapter 5 for a detailed discussion on the development of these values for design.

In force-based design, a force reduction factor, analogous to that used in ASCE for buildings, is used to reduce the elastic seismic force to a level intended to pre-clude collapse but permit ductile, inelastic behavior in selected elements. Table 4.2.5 summarizes these force reduction factors for three "operational categories."

In displacement-based design, displacement demands are determined at each substructure, typically by means of a linear, dynamic response spectrum analysis. The so-called equal displacement rule is assumed to estimate inelastic displacement demands as elastic displacement demands, with a modification factor for short period structures. The validity of the equal displacement rule has been shown in recent work to be valid, but over a limited range of periods. Ongoing research will likely provide for modifications to the equal displacement rule in AASHTO. Displacement capacity is estimated, either using empirical equations or by a pushover analysis,

---

TABLE 4.2.5
AASHTO Force Reduction Factors for Seismic Design

Substructure	Operational Category		
	Critical	Essential	Other
Wall type piers/bents	1.5	1.5	2.0
Reinforced concrete pile bents			
• Vertical piles	1.5	2.0	3.0
• Battered piles	1.5	1.5	2.0
Single column piers/bents	1.5	2.0	3.0
Steel or Composite Steel and Concrete Pile Bents			
• Vertical piles	1.5	3.5	5.0
• Battered piles	1.5	2.0	3.0
Multi-column piers/bents	1.5	3.5	5.0

and compared to displacement demands. The displacement capacity equations are presented here in Equation 4.2-6, for Seismic Design Category B, and Equation 4.2-7, for Seismic Design Category C. For Seismic Design Category D, the pushover analysis is required and no empirical equation is available in AASHTO.

$$\Delta_C^L = 0.12 H_o \left(-1.27 \ln(x) - 0.32 \right) \geq 0.12 H_o \qquad (4.2\text{-}6)$$

$$\Delta_C^L = 0.12 H_o \left(-2.32 \ln(x) - 1.22 \right) \geq 0.12 H_o \qquad (4.2\text{-}7)$$

$$x = \frac{\Lambda B_o}{H_o} \qquad (4.2\text{-}8)$$

- $\bullet$ = 1 for fixed-free column end conditions (cantilever)
- $\bullet$ = 2 for fixed-fixed column end conditions (rigid frame)
- $\bullet$ B_o is the column diameter for circular columns, or the dimension parallel to the direction in which displacement capacity is being calculated, for rectangular columns, ft
- $\bullet$ H_o is the clear column height, ft

Note that H_o and B_o are in ft, while the displacement capacity from the equations in inches.

The displacement capacity equations are intended for use with reinforced concrete column substructures with a minimum clear height of 15 ft. Attempts to use the equations for other situations will likely result in significant errors.

Once a substructure with adequate displacement capacity has been identified in the design process, the plastic shear of the substructure is determined, using overstrength and expected material properties. Connections from the superstructure to the substructure and elements below the column bases are designed for the application of the plastic shear at the center of gravity of the superstructure.

Example 4.2-5: Seismic Effects on Bridge Structures

Figure 4.2.9 is an elevation view of a proposed bridge. The bridge cross section (one-half of the dual bridge) is shown in Figure 4.2.10. Determine the transverse

FIGURE 4.2.9 Elevation of proposed bridge for seismic example.

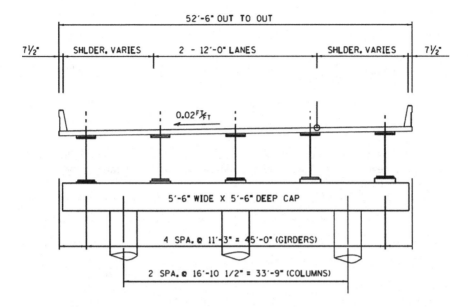

FIGURE 4.2.10   Cross section of bridge for seismic example.

design force from seismic effects using a force-based design. Determine the empirical transverse displacement capacity of Bent 1 for a displacement-based design. The bridge is an essential structure at a site in Seismic Design Category B with $S_{DS} = 0.55$ g and $S_{D1} = 0.30$ g. The total superstructure weight is 11.85 kips/ft, with two lanes of live load and an additional future overlay included in the weight calculation. Clear column height is 43.00 ft at Bent 1 and 33.64 ft at Bent 2. Column diameter is 54 inches. Span lengths are 195-ft, 230-ft, and 195-ft.

For a force-based design, AASHTO prescribes a seismic coefficient, C_{sm}, to be applied to the weight and divided by the force reduction factor to determine the design force. In the absence of a fundamental period, assume that $C_{sm} = S_{DS} = 0.55$, the maximum possible value.

The tributary weights at each pier are equal with the symmetric span arrangement. It has become somewhat customary to add, in addition to the cap weight, one-third of the column weight to the seismic weight at a given substructure:

$$W_{B1} = W_{B2} = 11.85 \text{ klf} \times \frac{1}{2}(195 \text{ ft} + 230 \text{ ft}) = 2{,}518 \text{ kips}$$

$$W_{CAP} = 5.5 \text{ ft} \times 5.5 \text{ ft} \times 55 \text{ ft} \times 0.150 \text{ kcf} = 250 \text{ kips}$$

$$W_{COLS} = 3 \times \frac{\pi 4.5^2}{4} \times 43 \text{ ft} \times 0.150 \text{ kcf} = 308 \text{ kips}$$

$$W_{EQ} = 2,518 + 250 + \frac{1}{3} \times 308 = 2,870 \text{ kips}$$

For a force-based design, the appropriate R-factor is 3.5 for the essential bridge:

$$F_{EQ} = \frac{0.55 \times 2,870}{3.5} = \frac{1,579}{3.5} = 451 \text{ kips}$$

For displacement-based design, use Equation 4.2-6 since the bridge is in Seismic Design Category B. Use $\Lambda = 2$ since the bent behaves as a rigid frame in the transverse direction.

$$x = \frac{2 \times 4.5}{43} = 0.2093$$

$$\Delta_C^L = 0.12(43)(-1.27 \ln(0.2093) - 0.32) = 8.60 \text{ inches} \geq 0.12(42) = 5.04 \text{ inches}$$

Section 4.7 presents a pushover analysis of this same bridge bent. The pushover analysis shows that the displacement capacity of the final designed column is about 13.59 inches. Pushover analysis will typically give a higher, more realistic estimate of displacement capacity compared to the empirical equations.

It should be noted that while this example is based on an actual proposed bridge, the true value for S_{DS} is 1.02 g and the true Seismic Design Category is D. Section 4.7, in addition to estimating the displacement capacity by pushover analysis, provides an estimate of the plastic shear, 709 kips. The overstrength plastic shear would thus be 1.2 × 709 = 851 kips. Applying a force-based design with strength reduction factor would give F_{EQ} = 1.02 × 2,870/3.5 = 836 kips. The force-based design in this case gives approximately the same design force level, but the provisions pay less attention to ductile design and detailing.

4.3   PRESTRESSED CONCRETE SUPERSTRUCTURES

Perhaps the most commonly used prestressing strand is 270 K low-relaxation strand. The tensile strength of this strand, f_{pu}, is 270 ksi, and the yield strength is typically 90% of the tensile strength. Typical 7-wire strand diameters are 0.5 inch (A = 0.153 in² per strand) and 0.6 inch (A = 0.217 in² per strand).

The initial pull on strands is typically 75% of the tensile strength of the strands. The AASHTO LRFD Bridge Design Specifications provide stress limits at various stages, in addition to Strength Limit State requirements for flexural and shear resistance.

AASHTO I-Beam and Bulb-Tee beams are common for prestressed superstructures. Approximate span limits for bulb-tee girders vary from location to location, but a value of 150 ft is typical. Some states design prestressed girders as simple spans. Others take advantage of continuity details to design the girders simple for dead load; continuous for live load and dead load applied after the deck strength has been attained. Washington State has used spliced girders to extend span limits considerably.

Rather than creating 3D models or 2D grillage models of superstructures, engineers frequently design the girders in a beam-type bridge using a continuous beam analysis. This simplified approach has worked well for decades and requires the determination of a so-called live load distribution factor. This factor may best be described as the number of traffic lanes for which each girder must be designed in the simplified line-girder analysis.

There are at least two methods available to the engineer in determining live load distribution factors for prestressed concrete bridge girders: (1) AASHTO equations and (2) the lever rule.

Parameters required in the AASHTO (AASHTO, AASHTO LRFD Bridge Design Specifications – 8th edition, 2017) equations for live load distribution factors are as follows:

- K_g: a stiffness parameter, $K_g = n[I + A(e_g)^2]$
- e_g: distance between the centers of gravity of the girder and the concrete deck, inches
- A: girder cross-sectional area, in^2
- I: girder moment of inertia about a horizontal axis, in^4
- n: modular ratio = E_S/E_C
- L: girder span, feet, taken as follows:
  - for positive moment, the span length
  - for negative moment, the average of adjacent spans
  - for shear, the span length
  - for interior reactions at continuous spans, the average of adjacent spans
- t_s: the concrete deck thickness, inches
- N_b: the number of girders in the cross section of the bridge
- S: girder spacing, center-to-center, ft
- d_e: distance from centerline of exterior web of exterior beam to face of curb, ft

Table 4.3.1 summarizes dimensions and properties for several popular I-girder types. For complete dimensional details, consult the PCI Bridge Design Manual (PCI, 2014).

Multi-span prestressed girders are often constructed to behave as a series of simple spans for noncomposite dead load and as continuous spans for composite dead load and live load. Noncomposite dead load consists of the girder self-weight, intermediate bracing required for stability, and the wet concrete deck. Composite dead loads typically include parapets, sidewalks, overlay, and utilities. Continuity is achieved by extending the prestressing strands from the end of the beams into a support diaphragm at intermediate piers. Details regarding research into this type of construction are available in the literature (Miller, Castrodale, Mirmiran, & Hastak).

Correction factors which increase shear live load distribution factors and reduce moment live load distribution factors may need to be applied for bridges with skewed

TABLE 4.3.1
AASHTO Prestressed Bridge Girder Properties

Girder	Depth, in	A, in^2	y_{bottom}, in	I, in^4	Wt., klf
Type I	28	276	12.59	22,750	0.287
Type II	36	369	15.83	50,980	0.384
Type III	45	560	20.27	125,390	0.583
Type IV	54	789	24.73	260,730	0.822
Type V	63	1,013	31.96	521,180	1.055
Type VI	72	1,085	36.38	733,320	1.130
Bulb-Tee-54	54	659	27.63	268,077	0.686
Bulb-Tee-63	63	713	32.12	392,638	0.743
Bulb-Tee-72	72	767	36.60	545,894	0.799

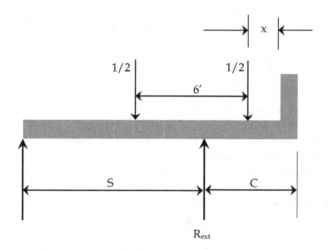

FIGURE 4.3.1   Lever rule live load distribution calculation.

substructures and are provided in the AASHTO Specifications (AASHTO, AASHTO LRFD Bridge Design Specifications – 8th edition, 2017).

For a typical bridge, the calculation of live load distribution factors for moment and shear in both interior and exterior girders is necessary. If spans are not equal, this adds even more distribution factors. One strategy is to determine the maximum of all distribution factors and design the entire bridge for that value, when this is possible and not economically disadvantageous.

The lever rule is based on the treatment of the exterior girder and the first interior girder providing a simple beam support of the deck with the overhang. Figure 4.3.1 is a schematic of the assumption inherent in the lever rule calculations. Each concentrated load represents one-half of a lane. The reaction at the exterior girder is taken as the distribution factor for purposes of the lever rule. The distance, x, is 2 ft for live load distribution factor calculations, and 1 ft for deck transverse reinforcement design. The cantilever dimension, C, is limited to $0.5S$ in the design specifications, but a value of no more than $0.40S$ is recommended.

The decreasing likelihood of simultaneous lanes being loaded in the most disadvantageous arrangement, as the number of loaded lanes increases, is incorporated into AASHTO live load design by the application of a multi-presence factor, m. The multi-presence factor is applied as follows:

- One loaded lane, $m = 1.20$
- Two loaded lanes, $m = 1.00$
- Three loaded lanes, $m = 0.85$
- Four or more loaded lanes, $m = 0.65$

The distribution factors in the AASHTO equations have the multi-presence factor, m, built in. There is no need to apply the multi-presence factor, m, to a distribution factor from the AASHTO equations. The symbol for the distribution factor in AASHTO is g. So the equations listed in AASHTO and here define mg. When a distribution factor is calculated using the lever rule, the multi-presence factor should be applied.

For prestressed I-girders, the live load distribution factors for moment in an interior girder are given by Equations 4.3-1 and 4.3-2. The equations apply for four or more girders with:

$$3.5 \leq S \leq 16.0$$

$$4.5 \leq t_s \leq 12.0$$

$$20 \leq L \leq 240$$

$$10,000 \leq K_g \leq 7,000,000$$

$$mg = 0.06 + \left(\frac{S}{14}\right)^{0.40}\left(\frac{S}{L}\right)^{0.30}\left(\frac{K_g}{12Lt_s^3}\right)^{0.10} \text{, one lane loaded} \qquad (4.3\text{-}1)$$

$$mg = 0.075 + \left(\frac{S}{9.5}\right)^{0.60}\left(\frac{S}{L}\right)^{0.20}\left(\frac{K_g}{12Lt_s^3}\right)^{0.10} \text{, multiple lanes loaded} \qquad (4.3\text{-}2)$$

With only three girders, the live load distribution factor for moment in an interior girder is to be taken as the lesser of that from Equations 4.3-1 and 4.3-2 or the lever rule.

For prestressed I-girders, the live load distribution factors for moment in an exterior girder are given by Equations 4.3-3 and 4.3-4. The equations apply for four or more girders with multiple lanes loaded and:

$$-1.0 \le d_e \le 5.5$$

$$e = 0.77 + \frac{d_e}{9.1} \tag{4.3-3}$$

$$mg = e\left(mg\right)_{interior} \tag{4.3-4}$$

With only one lane loaded, distribution factors for exterior girder moment are determined by the lever rule. With only three girders, the live load distribution factor for moment in an exterior girder is to be taken as the lesser of that from Equations 4.3-3 and 4.3-4 or the lever rule.

For prestressed I-girders, the live load distribution factors for shear in an interior girder are given by Equations 4.3-5 and 4.3-6. The equations apply for four or more girders with:

$$3.5 \le S \le 16.0$$

$$4.5 \le t_s \le 12.0$$

$$20 \le L \le 240$$

$$mg = 0.36 + \frac{S}{25}, \text{ one lane loaded} \tag{4.3-5}$$

$$mg = 0.2 + \frac{S}{12} - \left(\frac{S}{35}\right)^{2.0}, \text{ multiple lanes loaded} \tag{4.3-6}$$

With only three girders, the live load distribution factor for shear in an interior girder is to be calculated using the lever rule, regardless of the number of lanes loaded.

For prestressed I-girders, the live load distribution factors for shear in an exterior girder are given by Equations 4.3-7 and 4.3-8. The equations apply for four or more girders with multiple lanes loaded and:

$$-1.0 \le d_e \le 5.5$$

$$e = 0.60 + \frac{d_e}{10} \tag{4.3-7}$$

$$mg = e\left(mg\right)_{\text{interior}} \tag{4.3-8}$$

With only one lane loaded, or with only three girders, the live load distribution factor for shear in an exterior girder is to be calculated using the lever rule.

Example 4.3-1: PPC Girder Live Load Distribution Factor

Figure 4.3.2 depicts the proposed cross section for a three-span bridge. Span lengths are each 140 ft. Substructures are perpendicular to the bridge centerline (no skew). Twenty-eight-day concrete strengths are 10 ksi for the girder and 4 ksi for the deck. Determine the live load distribution factors for the following cases:

a) Moment in an interior girder
b) Moment in an exterior girder
c) Shear in an interior girder
d) Shear in an exterior girder

For moment in an interior girder, assume a 2-inch distance from the top of the girder centerline to the bottom of the deck in calculating e_g:

$$e_g = 72 - 36.6 + 2 + 8.25 \text{ inches} / 2 = 41.52 \text{ inches}$$

$$n = \sqrt{\frac{10}{4}} = 1.58$$

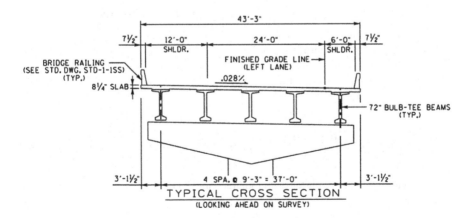

FIGURE 4.3.2 Bridge cross section for PPC girder live load distribution factor.

$$K_g = 1.58\left[545,894 + 767\left(41.52\right)^2\right] = 2,952,154 \text{ in}^4$$

$$mg = 0.06 + \left(\frac{9.25}{14}\right)^{0.40}\left(\frac{9.25}{140}\right)^{0.30}\left(\frac{2,952,154}{12\left(140\right)\left(8.25\right)^3}\right)^{0.10}$$

$$= 0.480 \text{ lanes, one lane loaded}$$

$$mg = 0.075 + \left(\frac{9.25}{9.5}\right)^{0.60}\left(\frac{9.25}{140}\right)^{0.20}\left(\frac{2,952,154}{12\left(140\right)\left(8.25\right)^3}\right)^{0.10}$$

$$= 0.716 \text{ lanes, multiple lanes loaded}$$

For moment in an exterior girder, with multiple lanes loaded:

$$d_e = 3.125 \text{ ft} - 0.625 \text{ ft} = 2.500 \text{ ft}$$

$$e = 0.77 + \frac{2.500}{9.1} = 1.045$$

$$mg = 1.045 \times 0.716 = 0.748 \text{ lanes}$$

For moment in an exterior girder, with a single lane loaded, use the lever rule, sum moments about the interior girder to find the exterior girder reaction. Apply a multi-presence factor $m = 1.20$ since only one lane is loaded. See Figure 4.3.3.

$$mg = R_{ext} = 1.20\left[\frac{1}{2}\left(3.75 + 9.75\right)/9.25\right] = 0.876 \text{ lanes}$$

For shear in an interior girder:

$$mg = 0.36 + \frac{9.25}{25} = 0.730 \text{ lanes, one lane loaded}$$

$$mg = 0.2 + \frac{9.25}{12} - \left(\frac{9.25}{35}\right)^{2.0} = 0.901 \text{ lanes, multiple lanes loaded}$$

For shear in an exterior girder, with multiple lanes loaded:

$$e = 0.60 + \frac{2.500}{10} = 0.850$$

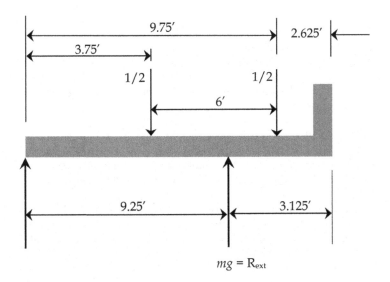

$$mg = R_{ext}$$

FIGURE 4.3.3 Lever rule calculation example.

TABLE 4.3.2
Live Load Distribution Factor Example Summary

Condition	Mg	Controlling Case
Moment, interior girder	0.716 lanes/girder	Multiple lanes loaded
Moment, exterior girder	0.876 lanes/girder	Single lane loaded
Shear, interior girder	0.901 lanes/girder	Multiple lanes loaded
Shear, exterior girder	0.876 lanes/girder	Single lane loaded

$$mg = 0.850 \times 0.901 = 0.766 \text{ lanes}$$

For shear in an exterior girder, with a single lane loaded, the lever rule with multipresence factor incorporated has already been shown to give $mg = 0.876$ lanes per girder.

Table 4.3.2 summarizes the live load distribution factors and controlling cases for the bridge girders.

To be clear, live load distribution factors permit girder design to be completed using a line-girder analysis (i.e., a single continuous beam model) rather than a three-dimensional or grillage model. This greatly simplifies girder design and the interpretation of results. To determine design moments in an exterior girder of the subject bridge, the engineer could use a three-span continuous beam model and apply 87.6% of a full HL-93 live loading (lane plus truck) to the model. Of course, the appropriate impact factors and load factors would then need to be applied for each applicable limit state, and load combination. These calculations are done internally in modern bridge design software. However, it is important that the engineer understand the concept of a live load distribution factor in performing sanity checks of existing software or in writing personal programs.

Recall also that the multi-presence factor, m, does not apply for the Fatigue Limit State, since this limit state is by definition a single lane with a single design truck. For fatigue calculations, the distribution factors for a single lane loaded, without multi-presence, are as follows:

- 0.480 / 1.20 = 0.400 lanes per girder, Moment in Interior Girder
- 0.876 / 1.20 = 0.730 lanes per girder, Moment in Exterior Girder
- 0.730 / 1.20 = 0.608 lanes per girder, Shear in Interior Girder
- 0.876 / 1.20 – 0.730 lanes per girder, Shear in Exterior Girder

A completed structure used as the basis for this example is shown in Figure 4.3.4.

Of particular interest for the current discussion is the nominal moment resistance in positive bending (top of the deck in compression). For this condition, the stress in the strand at nominal resistance is given by Equation 4.3-9. In this equation, d_p is the distance from the top of the deck to the centroid of the prestressing strands. The tensile strength of the strand is f_{pu}, typically 270 ksi, though other grades are available. The distance from the top of the deck to the neutral axis is c and is given by Equation 4.3-12 if the stress block depth is less than or equal to the deck thickness. The coefficient, k, is given by Equation 4.3-10 and is equal to 0.28 for low-relaxation strand. For prestressed girders composite with concrete deck

FIGURE 4.3.4   State Route 52 over Branch in Clay County, TN.

having a compressive strength of f'_c, and with compressive stress block depth, $a = \beta_1 c$, less than or equal to the deck thickness, the nominal positive flexural resistance is given by Equation 4.3-11. The coefficient, β_1, is 0.85 for a concrete strength of 4 ksi or less, 0.65 for a concrete strength of 8 ksi or more, and a linear interpolation is used for intermediate concrete strength values. The coefficient, α_1, is 0.85 for a concrete strength of 10 ksi or less, 0.75 for a concrete strength of 15 ksi or more, and a linear interpolation is used for intermediate concrete strength values. A_{ps} is the total area of all developed prestressing strands.

$$f_{ps} = f_{pu}\left(1 - \frac{kc}{d_p}\right) \tag{4.3-9}$$

$$k = 2\left(1.04 - \frac{f_{py}}{f_{pu}}\right) \tag{4.3-10}$$

$$M_n = A_{ps}f_{ps}\left(d_p - \frac{a}{2}\right) \tag{4.3-11}$$

$$c = \frac{A_{ps}f_{pu}}{\alpha_1 f'_c \beta_1 b + k A_{ps}f_{pu} / d_p} \tag{4.3-12}$$

Example 4.3-2: Positive Moment Resistance of a Composite Prestressed Girder

Figure 4.3.5 depicts the strand pattern used for the BT-72 girders used in the bridge shown in Figure 4.3.4. The left sketch shows the mid-span strand pattern and the right sketch shows the girder end strand pattern (draped, or harped, strands were used in the design). The girder length is 138.25 ft and the drape point for the strands is 55 ft from each beam end. Determine the initial pull on the strands, the hold-down force at the drape point, and estimate the composite, positive moment resistance of the girder based on the mid-span strand pattern (the point of maximum positive moment). Strands are 0.6-inch diameter, low-relaxation, with $f_{pu} = 270$ ksi. Strand area is 0.217 in² per strand. The deck thickness is 8.25 inches and the gap between the top of the BT-72 girder and the bottom of the concrete deck is 2 inches.

The centroid of the 48 strands first must be located, relative to the bottom of the BT-72 girder.

$$\bar{y}_{ps} = \frac{12(2+4)+4(6)+2(8+10+12+14+16+28+30+32+34+36)}{48} = 11.17 \text{ inches}$$

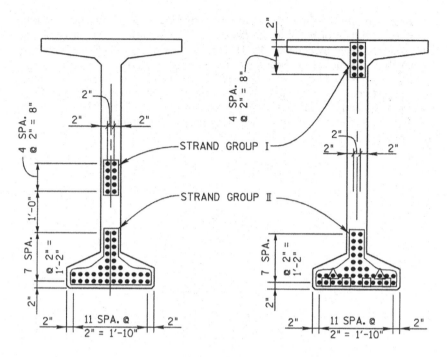

FIGURE 4.3.5   Strand Pattern for prestressed concrete BT-72 girder.

$$A_{ps} = 48 \times 0.217 = 10.416 \text{ in}^2$$

The initial pull force, based on 75% of the specified tensile strength of the strands, is easily determined.

$$P_{init} = 48 \text{ strands} \times 0.75 \times 270 \text{ ksi} \times 0.217 \text{ in}^2 = 2{,}109 \text{ kips}$$

Assume the compressive stress block depth is less than the deck thickness and estimate the stress block depth, c. Since the deck, not the beam, is in compression, under this condition, use $\alpha_1 = 0.85$ and $\beta_1 = 0.85$.

$$d_p = 72 + 2 + 8.25 - 11.17 = 71.08 \text{ inches}$$

$$c = \frac{10.416(270)}{0.85(4)(0.85)(111) + 0.28(10.416)(270)/71.08} = 8.47 \text{ inches}$$

$$a = 0.85 \times 8.47 = 7.20 \text{ inches}$$

$$f_{ps} = 270\left(1 - \frac{0.28 \times 8.47}{71.08}\right) = 261 \text{ ksi}$$

$$M_n = (10.416)(261)\left(71.08 - \frac{7.2}{2}\right) = 183,449 \text{ in}\cdot\text{k} = 15,287 \text{ ft}\cdot\text{k}$$

From the strand pattern sketch, note that the strands being draped are raised 34 inches = 2.83 ft over the distance of 55 ft. The angle of the drape may be used to estimate the hold-down force. Only 10 of the strands, indicated in the sketch, are draped.

$$\left(P_{init}\right)_{draped} = 10 \times 0.217 \times 0.75 \times 270 = 439 \text{ kips}$$

The hold-down force is the vertical component of the draped strand initial pull.

$$F_{HD} = 439 \cdot \frac{2.83}{\sqrt{2.83^2 + 55^2}} = 22.5 \text{ kips total } \left(2.25 \text{ kips per strand}\right)$$

Note that the stress block depth, a, is in fact less than the deck thickness, but the distance to the neutral axis, c, is greater than the deck thickness. The error resulting is minimal. A section analysis program, Response 2000, gives a nominal moment resistance of 15,497 ft·kips. The estimate from the provided equations is very close to the more accurate solution from a section analysis.

This example is intended to provide the reader with a basic idea behind pre-stressed concrete girder calculations, not to give a comprehensive review of design requirements.

For a complete set of criteria for the design of prestressed concrete girders, the reader is directed to Chapter 5 of AASHTO LRFD Bridge Design Specifications.

4.4   STEEL GIRDER SUPERSTRUCTURES

Welded steel plate girders are more common than rolled steel beams for bridge structures.

With regard to live load distribution factors, those presented in Section 4.3 for prestressed girders are also applicable to steel girders. However, with steel girder bridges, the cross section properties are unknown before design of the girder has been completed. AASHTO provides for an approximate distribution factor to be cal-culated using preliminary values for the following when steel girders are proposed:

$$\left(\frac{K_g}{12Lt_s^3}\right)^{0.10} = 1.02$$

Final design should include calculation of the stiffness parameter, K_g, and the actual distribution factor value from the equations once the cross section properties are known.

The "rigid cross section" method of live load distribution is often necessary for steel girder bridges as well. AASHTO 4.6.2.2.2d specifies that for exterior girders in beam-slab bridges with diaphragms or cross-frames, the live load distribution factor should also be computed using this method. Therefore, even with concrete girders, if rigid intermediate diaphragms are present in the bridge, the method should be applied. Implicit in the method is the assumption that the cross section rotates as a rigid body. The "rigid cross section" method live load distribution factor is defined by Equation 4.4-1. N_L is the number of loaded lanes. N_B is the number of beams in the cross section. The distance from the centerline of the bridge to the center of a lane is the eccentricity, e, for that lane. The distance from the centerline of the bridge to a girder centerline is the parameter, X, for that girder.

$$mg = m \left[\frac{N_L}{N_B} + \frac{X_{ext} \sum e}{\sum X^2} \right]$$

(4.4-1)

Example 4.4-1: Rigid Cross Section Method for Live Load Distribution

Figure 4.4.1 depicts the proposed cross section for a bridge on Forrester Road in Obion County, TN. Compute the exterior girder live load distribution factor using (a) the lever rule and (b) the rigid cross section method.

Only two lanes will physically fit on the bridge, so consider both one-lane and two-lane cases with appropriate multi-presence factors applied.

For the lever rule, the outermost wheel is 31.25 ft/2 − 7.5 inches/12 − 2 ft = 13.00 ft from the centerline of the bridge, which, in this case, is also the first interior girder for lever rule calculations. The adjacent wheel is 13 ft − 6 ft = 7 ft from

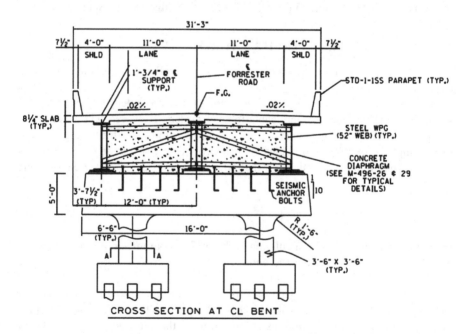

FIGURE 4.4.1   Forrester Road Bridge cross section.

the centerline of the bridge. The single lane, lever rule method, live load distribution factor is thus:

$$mg = 1.20\left[\frac{1}{2}\cdot\frac{(13+7)}{12}\right] = 1.000 \text{ lanes per girder}$$

For the rigid cross section method, the eccentricity of the outermost lane is $e_1 = 13$ ft – 3 ft = 10 ft. The eccentricity of the adjacent lane is $e_2 = e_1 - 12$ ft = 10 ft – 12 ft = –2 ft.

With only the outermost lane loaded, the distribution factor by rigid cross section calculation gives:

$$\sum X^2 = (-12)^2 + (0)^2 + (12)^2 = 288 \text{ ft}^2$$

$$\sum e = 10 \text{ ft}$$

$$mg = 1.20\left[\frac{1}{3}+\frac{12\times10}{288}\right] = 0.900 \text{ lanes per girder}$$

With both lanes loaded:

$$\sum e = 10 + (-2) = 8 \text{ ft}$$

$$mg = 1.00\left[\frac{2}{3}+\frac{12\times8}{288}\right] = 1.000 \text{ lanes per girder}$$

Some of the commonly used bridge steels are summarized in Table 4.4.1. Table 4.4.2 lists properties for commonly used high-strength bolts in bridge structures.

TABLE 4.4.1
Bridge Steel Properties (AASHTO M270/ASTM A709)

Grade	36	50	50 W	HPS 50 W	HPS 70 W	HPS 100 W	HPS 100 W
Shapes	All	All	All	NA	NA	NA	NA
Plates	≤4″	≤4″	≤4″	≤4″	≤4″	≤2.5″	2.5″–4″
F_y, ksi	36	50	50	50	70	100	90
F_u, ksi	58	65	70	70	85	110	100

TABLE 4.4.2
Anchor Rod and High-Strength Bolt Properties

ASTM Designation	F_y, ksi	F_u, ksi	Category
F3125 Grade A325	–	120	High-strength bolt
F3125 Grade A490	–	150	High-strength bolt
F1554 Grade 36	36	58	Anchor rod
F1554 Grade 55	55	75	Anchor rod
F1554 Grade 105	105	125	Anchor rod

Five aspects of steel girder design that receive special attention in this text are (a) field splice design, (b) fatigue design, (c) stability design, (d) flexural resistance of I-girders, and (e) shear resistance of I-girders.

4.4.1 FIELD SPLICE DESIGN

Shipping section lengths for steel girders are limited, typically to 120–160 ft, depending on the route to be taken. Longer lengths may be shipped via barge when available. Steel girder spans for modern bridges are typically longer than the shipping limit, and a filed splice is often required. Section 6.13.6 in the AASHTO LRFD Bridge Design Specifications provides detailed requirements for field splices. Additional literature provides excellent guidance and design examples (Grubb, Frank, & Ocel, 2018).

Figure 4.4.2 shows an elevation and section of two girder field sections requiring splicing and identifies the nine primary components of an I-girder field splice:

1. Top flange, outer splice plate
2. Top flange, inner splice plates
3. Web splice plates
4. Bottom flange, outer splice plate
5. Bottom flange, inner splice plates
6. Top flange bolts
7. Web bolts
8. Bottom flange bolts
9. Filler plates (required when flange thickness is not the same on both sections)

Some of the basic requirements for field splice design require only a few simple calculations:

- The sum of splice plate areas for a flange splice must equal or exceed the area of the controlling flange being spliced.
- The sum of areas for web splice plates must equal or exceed the web area.

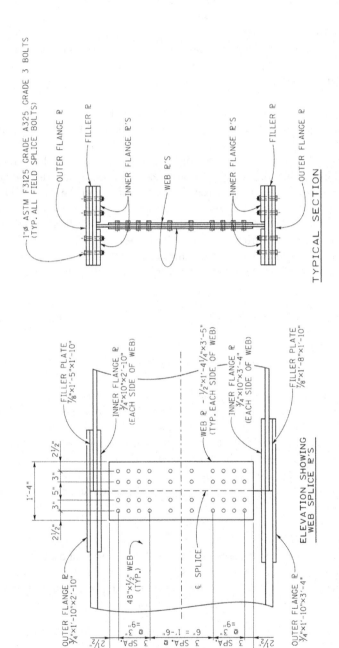

FIGURE 4.4.2 Steel Bridge girder splice elevation and section.

- The ratio of inner splice plate area to outer splice plate area must be between 0.90 and 1.10 in order for the assumption of equal load distribution to inner and outer plates to be valid.
- Fatigue need not be checked for bolted field splice designs because the current version of the Specification provisions is designed to preclude fatigue failure in the splice.

Detailed provisions require calculation of the girder moments for the Service II Limit State during deck casting and for the Strength I Limit State for the completed, in-service bridge. Equations 4.4-2 and 4.4-3 summarize requirements for the flange splice. P_{fy} is the design yield resistance of the flange splice and the force for which the splice plates and bolts must be designed. Resistance factors for rupture (ϕ_u) and yield (ϕ_y) are 0.80 and 0.95, respectively, in the 8th edition of the AASHTO LRFD Bridge Design Specifications. F_{uf} and F_{yf} are tensile and yield strength values for the flange plate. A_n is the net area and A_g is the gross area of the flange plate. For net area calculations, the standard hole size is 1/16 inches larger than the bolt diameter for bolts up to and including 7/8-inch diameter, and 1/8 inches larger than the bolt diameter for bolts 1-inch in diameter and larger.

$$P_{fy} = F_{yf} A_e \qquad (4.4\text{-}2)$$

$$A_e = \frac{\phi_u}{\phi_y} \cdot \frac{F_{uf}}{F_{yf}} \cdot A_n \le A_g \qquad (4.4\text{-}3)$$

AASHTO provides a means of calculating the moment resistance provided by the flanges. If this moment is less than the Strength I Limit State moment at the splice, then the web must be designed to carry the excess moment in addition to the shear. For composite sections in positive bending (the bottom flange being in tension), the moment resistance provided by the flanges is determined from Equation 4.4-4. The moment arm, A_1, is the distance from mid-thickness of the bottom flange to mid-thickness of the concrete deck, and $(P_{fy})_{bottom}$ is the design force for the bottom flange.

$$M_{fl} = \left(P_{fy}\right)_{bottom} \cdot A_1 \qquad (4.4\text{-}4)$$

For composite sections in negative flexure and for non-composite sections, the moment resistance provided by the flanges is given by Equation 4.4-5. The moment arm, A_2, is the distance between mid-thickness of the top and bottom flanges, and $(P_{fy})_{min}$ is the smaller of the design forces for the top and bottom flanges.

$$M_{fl} = \left(P_{fy}\right)_{min} \cdot A_2 \qquad (4.4\text{-}5)$$

When the moment resistance provided by the flanges is less than the Strength Limit State moment, the web splice must be designed for a horizontal component of force, H_w, acting simultaneously with a vertical component of force, V_r. Equations 4.4-6 and 4.4-7 provide a means of determining H_w for sections in positive moment and sections in negative moment, respectively. The moment arm, A_3, is the distance from mid-height of the web to mid-thickness of the concrete deck. The moment arm, A_4, is the distance from mid-height of the web to mid-thickness of the flange having the larger design force.

$$H_w = \frac{M_u - M_{fl}}{A_3} \qquad (4.4\text{-}6)$$

$$H_w = \frac{M_u - M_{fl}}{A_4} \qquad (4.4\text{-}7)$$

The minimum spacing of standard holes is three times the bolt diameter, measured center-to-center of bolts. The maximum spacing of bolts, to prevent moisture intrusion, is dependent on the thickness (t) of the thinner plate, and is given by Equation 4.4-8.

$$s \le 4 + 4t \le 7.0 \text{ inches} \qquad (4.4\text{-}8)$$

Bolt shear resistance at the Strength Limit State is given by Equation 4.4-9. A penalty factor of 0.83 is to be applied to the design shear resistance whenever the connection is longer than 38 inches. This penalty applies to tension lap-splices. It should not be applied to the web splice design. F_{ub} is the tensile strength of the bolts, A_b is the nominal cross-sectional area of the bolt, and N_s is the number of shear planes.

$$\phi R_n = 0.80 \times \begin{cases} 0.56 F_{ub} A_b N_s, & \text{threads excluded from shear planes} \\ 0.45 F_{ub} A_b N_s, & \text{threads not excluded from shear planes} \end{cases} \qquad (4.4\text{-}9)$$

Web splices typically will have threads included in the shear planes since both the splice plates and the web plate are relatively thin. For ¾-inch and thicker flange plates, it is usually possible to exclude threads from the shear planes. Such an assumption, if made, should be noted on the design plans.

When filler plates ¼-inch and thicker are used in a field splice, an additional penalty factor is applied to the element containing the filler plates. The penalty factor is given by Equation 4.4-10. The parameter, γ, is the sum of the filler plate areas (A_f) divided by the smaller of either (a) the connected plate area or (b) the sum of the splice plate areas (A_P).

$$R = \frac{1 + \gamma}{1 + 2\gamma} \qquad (4.4\text{-}10)$$

Before proceeding to a field splice example, note that it is necessary to determine the design shear resistance for the web in order to determine the design forces for the web splice. Equation 4.4-11 is the design shear resistance for an unstiffened web ($k = 5$) or for a stiffened web in an end panel of an I-girder. Equation 4.4-12 gives the design resistance of a stiffened web in an interior panel of an I-girder. Equation 4.4-15 gives the design force, R_{web}, for the web splice plates and bolts. F_{yw} is the yield strength of the web plate, D is the web depth, t_w is the web thickness, E is Young's modulus for steel (29,000 ksi), and d_o is the transverse stiffener spacing.

$$\phi V_n = 1.0 \times C \times \left[0.58 F_{yw} D t_w \right] \tag{4.4-11}$$

$$\phi V_n = 1.0 \times \left[C + \frac{0.87(1-C)}{\sqrt{1+(d_o/D)^2}} \right] \times \left[0.58 F_{yw} D t_w \right] \tag{4.4-12}$$

$$k = 5 + \frac{5}{(d_o/D)^2} \tag{4.4-13}$$

$$C = \begin{cases} 1.0, \text{ if } \dfrac{D}{t_w} \le 1.12 \sqrt{\dfrac{Ek}{F_{yw}}} \\[2em] \dfrac{1.12}{\left(D/t_w\right)} \sqrt{\dfrac{Ek}{F_{yw}}}, \text{ if } 1.12 \sqrt{\dfrac{Ek}{F_{yw}}} < \dfrac{D}{t_w} < 1.40 \sqrt{\dfrac{Ek}{F_{yw}}} \\[2em] \dfrac{1.57}{\left(D/t_w\right)^2} \times \left(\dfrac{Ek}{F_{yw}} \right), \text{ if } \dfrac{D}{t_w} \ge 1.40 \sqrt{\dfrac{Ek}{F_{yw}}} \end{cases} \tag{4.4-14}$$

$$R_{web} = \sqrt{\left(\phi V_n\right)^2 + \left(H_w\right)^2} \tag{4.4-15}$$

Equation 4.4-12 for stiffened interior panels typically applies, but is limited to those cases for which Equation 4.4-16 is satisfied. If Equation 4.4-16 is not satisfied, then the design shear resistance for interior stiffened panels is reduced to that given by Equation 4.4-17. The compression flange dimensions are b_{fc} and t_{fc}, while the tension flange dimensions are b_{ft} and t_{ft}.

$$\frac{2 D t_w}{b_{fc} t_{fc} + b_{ft} t_{ft}} \le 2.5 \tag{4.4-16}$$

$$\phi V_n = 1.0 \times \left[C + \frac{0.87(1-C)}{\sqrt{1+(d_o/D)^2} + (d_o/D)} \right] \times \left[0.58 F_{yw} D t_w \right] \tag{4.4-17}$$

Example 4.4-2: Steel I-Girder Field Splice Design

For the field splice depicted in Figure 4.4.2, assess the adequacy of the splice design. All material for both the girder and the splice plates is A709 Grade 50W ($F_y = 50$ ksi, $F_u = 70$ ksi). Bolts are 1-inch diameter ASTM F3125 Grade A325 with threads excluded from shear planes for the flanges and included in the shear planes for the web. The distance from the top of the web to the top of the 8.25-inch thick concrete deck is 11.75 inches. The Strength Limit State moment from structural analysis at the splice centerline is −5,150 ft·kips. The Service II deck-casting moment is −3,410 ft·kips. The negative signs on moment simply indicate that the top flange is in tension (a negative moment section by bridge convention). The transverse stiffener spacing is 12 ft. A field splice is an interior panel for shear resistance calculations.

The top flange plate is 22 inches × 1.125 inches right of the splice centerline, and 22 inches × 2 inches left of the splice centerline. The bottom flange plate is 22 inches × 1.125 inches right of the splice centerline, and 22 inches × 2 inches left of the splice centerline. Notice that a doubly symmetric girder section was employed in this design.

For either the top or the bottom flange splice, check the area requirement for the splice plates:

$$A_{flange} = 22 \times 1.125 = 24.75 \text{ in}^2$$

$$A_{inner} + A_{outer} = 2(10 \times 0.75) + 22 \times 0.75 = 31.50 \text{ in}^2 > 24.75, \text{ OK}$$

Check the ratio of inner plate to outer plate area:

$$0.90 < \frac{A_{inner}}{A_{outer}} = \frac{2 \times 10 \times 0.75}{22 \times 0.75} = 0.909 < 1.10, \text{ OK}$$

Determine the net area, A_n, of the flange plate, noticing that there are four rows of bolts in each flange:

$$A_n = [22 - 4 \times 1.125]1.125 = 19.69 \text{ in}^2$$

Determine the effective area and the flange splice design force:

$$A_e = \frac{0.80}{0.95} \cdot \frac{70}{50} \cdot 19.69 = 23.21 < 24.75 \rightarrow A_e = 23.21 \text{ in}^2$$

$$P_{fy} = 50 \times 23.21 = 1,160 \text{ kips}$$

Determine the design resistance in shear for the 1-inch diameter bolts in the flanges. Note that since both inner and outer splice plates are typically used in field splices for bridge girders, as is the case in this example, the number of shear planes $N_s = 2$. The nominal area for a 1-inch diameter bolt is 0.785 in². The tensile strength for F3125 Grade A325 bolts is 120 ksi. Apply the penalty factor for filler plates on the top flange bolt resistance, since 7/8-inch thick filler plates are used on both the top and bottom flange splices.

$$A_f = 22 \times 0.875 = 19.25 \text{ in}^2$$

$$A_P = Min(24.75, 31.50) = 24.75 \text{ in}^2$$

$$\gamma = \frac{19.25}{24.75} = 0.778$$

$$R = \frac{1+0.778}{1+2 \times 0.778} = 0.696$$

$$\phi R_n = 0.696 \times 0.80 \times 0.56 \times 120 \times 0.785 \times 2 = 58.7 \text{ kips per bolt}$$

Determine the design resistance in shear for the 1-inch diameter bolts in the web.

$$\phi R_n = 0.80 \times 0.45 \times 120 \times 0.785 \times 2 = 67.8 \text{ kips per bolt}$$

Determine the number of bolts required in each flange splice on each side of the splice centerline. To determine this, apply the required penalty factor since 7/8-inch thick filler plates are used.

$$n_b = \frac{1,160}{58.7} = 19.8 \text{ bolts} \rightarrow \text{use 5 rows, 4 bolts per row} = 20 \text{ bolts}$$

The total number of flange bolts for the splice in one girder will be 4 × 20 = 80 bolts (20 top left, 20 top right, 20 bottom left, 20 bottom right).

Determine the moment resistance of the flanges, noting that this is a composite section in negative flexure.

$$A_2 = 48 + 1.125 = 49.125 \text{ inches}$$

$$M_{fl} = 1,160 \cdot 49.125 \div 12 = 4,749 \text{ ft} \cdot \text{kips}$$

Since the flexural resistance of the flanges is less than the Strength Limit State moment of 5,150 ft·kips, the web splice design must include the horizontal force, H_w in addition to the vertical force, V_r.

$$A_4 = \frac{1}{2}(48) + \frac{1}{2}(1.125) = 24.56 \text{ inches}$$

$$H_w = \frac{5,150 \text{ ft}\cdot\text{k} - 4,749 \text{ ft}\cdot\text{k}}{24.56 \text{ inches}/12} = 196 \text{ kips}$$

$$d_o / D = \frac{12 \text{ft}}{4 \text{ft}} = 3$$

$$k = 5 + \frac{5}{3^2} = 5.556$$

$$\frac{D}{t_w} = \frac{48}{0.50} = 96 > 1.40\sqrt{\frac{Ek}{F_{yw}}} = 1.40\sqrt{\frac{29,000 \times 5.556}{50}} = 79.5$$

$$C = \frac{1.57}{(96)^2} \times \left(\frac{29,000 \times 5.556}{50} \right) = 0.549$$

$$\frac{2Dt_w}{b_{fc}t_{fc} + b_{ft}t_{ft}} = \frac{2 \times 48 \times 0.50}{22 \times 1.125 + 22 \times 1.125} = 0.97 \le 2.5$$

$$V_r = \phi V_n = 1.0 \times \left[0.549 + \frac{0.87(0.451)}{\sqrt{1+(3)^2}} \right] \times [0.58 \times 50 \times 48 \times 0.50] = 468 \text{ kips}$$

$$R_{web} = \sqrt{(468)^2 + (196)^2} = 507 \text{ kips}$$

Determine the number of web bolts required.

$$n_b = \frac{507}{67.8} = 7.5 \text{ bolts per side of centerline}$$

The 20 bolts used in the design is more than adequate.

This design example is not exhaustive, but serves to illustrate the concept and initial calculations for an I-girder bolted field splice. Additional considerations that must be addressed include bolt slip at the Service Limit State, tensile rupture of the splice plates, tensile yield of the splice plates, block shear of the splice plates, and bearing and tear-out resistance at the holes. Refer to the AASHTO LRFD Bridge Design Specifications, Section 6.13, and to the referenced literature for complete design requirements.

4.4.2 FATIGUE DESIGN

Design for fatigue is important for most any bridge structure, perhaps particularly so for steel bridge superstructures given the prevalence of welding in such structures. Fatigue is of concern whenever a cyclic stress state exists, with one extreme being tension. Fatigue is related to the initiation and propagation of cracks, often at stress levels well below the yield strength of the material. Welding introduces micro-cracks into the base metal being welded.

AASHTO prescribes two Fatigue Limit States: Infinite Life (Fatigue Limit State I) and Finite Life (Fatigue II). The impact factor (IM) is 15% for both Fatigue Limit States. Load factors are 1.75 for Fatigue I and 0.80 for Fatigue II. Recall from previous discussion that the multi-presence factor, m, does not apply for fatigue loading. The fatigue loading is defined as one fatigue truck in one lane. The only difference between the fatigue truck and the HL-93 design truck (see Figure 4.1.1) is the spacing between the two 32 kip axles. For the design truck, the spacing varies between 14 and 30 ft to produce the most severe effect. For the fatigue truck, the spacing is constant at 30 ft.

The basis for fatigue design in the AASHTO LRFD Bridge Design Specifications is $ADTT_{SL}$, the average daily truck traffic in a single lane. When the $ADTT_{SL}$ is not available, AASHTO provides a means for estimating it. The estimate is presented in Equation 4.4-18 here, as a function of either the ADT (average daily traffic) or the $ADTT$ (average daily truck traffic).

$$\left(ADTT\right)_{SL} = p \times ADTT = p \times fr \times ADT \qquad (4.4\text{-}18)$$

$$p = \begin{cases} 1.0, \text{ if only one lane is available to trucks} \\ 0.85, \text{ if two lanes are available to trucks} \\ 0.80, \text{ if three or more lanes are available to tricks} \end{cases} \qquad (4.4\text{-}19)$$

$$fr = \begin{cases} 0.20, \text{ for rural interstates} \\ 0.15, \text{ for urban interstates} \\ 0.15, \text{ for rural other} \\ 0.10, \text{ for urban other} \end{cases} \qquad (4.4\text{-}20)$$

AASHTO fatigue design depends on stress ranges and the number of cycles of stress range in steel elements caused by the passage of the fatigue truck across the bridge. In some instances, the passage of the fatigue truck across the bridge causes more than a single cycle of stress range. For computations, AASHTO specifies $n = 1.5$ cycles per truck passage for continuous beams near interior supports and $n = 1.0$ cycles per truck passage otherwise in beam bridges.

Eight fatigue detail categories are assigned for fatigue design. Category A is the most favorable condition and Category E is the least favorable condition. Each fatigue category has associated parameters, A and ΔF_{TH}. A is a constant used in determining the number of cycles to failure for the Fatigue II Limit State. ΔF_{TH} is the fatigue threshold, the stress below which fatigue failure presumably will not occur regardless of the number of cycles. ΔF_{TH} is used in Fatigue I, Infinite Life fatigue design.

Table 4.4.3 summarizes the parameters for the eight fatigue categories in AASHTO.

For Fatigue II, Finite Life fatigue, the number of cycles of stress is computed according to Equation 4.4-21, which has an explicit assumption of a 75-year design life. The design fatigue resistance, $(\Delta F)_n$, for the Fatigue II Limit State is given by Equation 4.4-22.

$$N = 365(75)(n)(ADTT)_{SL} \tag{4.4 21}$$

$$(\Delta F)_n = \left(\frac{A}{N}\right)^{\frac{1}{3}} \tag{4.4-22}$$

For Fatigue I, Infinite Life fatigue design, the design fatigue resistance $(\Delta F)_n$ is taken equal to the fatigue threshold, ΔF_{TH}.

TABLE 4.4.3

AASHTO Fatigue Parameters, A and ΔF_{TH}

Detail Category	A (ksi³)	ΔF_{TH} (ksi)
A	250×10^8	24.0
B	120×10^8	16.0
B'	61×10^8	12.0
C	44×10^8	10.0
C'	44×10^8	12.0
D	22×10^8	7.0
E	11×10^8	4.5
E'	3.9×10^8	2.6

Recall from Section 4.2 that the load factors are $\gamma = 1.75$ for Fatigue I $\gamma =$ and 0.80 for Fatigue II limit states.

Example 4.4-3: Fatigue Resistance Calculations

The top flange of a welded plate girder near the interior support of a continuous span bridge has shear studs. The ADT is 8,062 vehicles per day. The bridge is on a rural Interstate with four lanes available to truck traffic. Determine the permissible stress range for fatigue considerations.

For fatigue design, the welding of studs on the top flange of a steel girder produces a Category C detail. For Category C, the fatigue constant $A = 44 \times 10^8$ ksi^3 and $\Delta F_{TH} = 10$ ksi.

Consider first the Finite Life, Fatigue II Limit State.

$$ADTT_{SL} = 8,062 \times 0.80 \times 0.20 = 1,290 \text{ trucks per lane per day}$$

$$N = 365(75)(1.5)(1,290) = 52,970,625 \text{ cycles}$$

$$(\Delta F)_n = \left(\frac{44 \times 10^8}{52,970,625} \right)^{\frac{1}{3}} = 4.36 \text{ ksi}$$

Determine the permissible stress range, Δf, from the passage of the fatigue truck across the bridge (with 15% added for impact since $IM = 0.15$ for the Fatigue Limit State), incorporating the appropriate load factor for the Fatigue II Limit State.

$$0.80(\Delta f) \le 4.36 \rightarrow \Delta f \le 5.45 \text{ ksi}$$

Next, consider the Infinite Life, Fatigue I Limit State.

$$1.75(\Delta f) \le 10.0 \rightarrow \Delta f \le 5.71 \text{ ksi}$$

So if the engineer designs the girder such that the stress range in the top flange at the point in question is no more than 5.71 ksi, infinite fatigue life could be expected for the detail in question. Infinite fatigue life controls the design in this situation. There is no need to limit the stress range to 5.45 ksi indicated in the Fatigue II analysis.

In cases where Fatigue II produces a higher Δf than does Fatigue I, the engineer has two options: limit Δf to that indicated for Fatigue I and produce a design with infinite fatigue life, or limit the stress to the higher Δf from Fatigue II and produce a finite fatigue life design. Some engineers choose to design for infinite life always, which does seem reasonable given uncertainties in the calculation of number of cycles.

Fatigue failures have occurred in bridge superstructures. Figure 4.4.3 illustrates such an example: Interstate 95 over the Brandywine River in Delaware. This photo serves to stress the point that fatigue in bridge structures is not just an academic problem.

4.4.3 STABILITY DESIGN

Structural stability at all phases of a project, and in particular during construction of steel girder bridges, is a complex problem and inadequate attention to stability has resulted in structural collapse during construction. Examples of such failures include:

- State Route 69 over the Tennessee River in Clifton, TN, 1995 (Figure 4.4.4)
- The Westgate Bridge, Melbourne, Australia, 1970
- Quebec Bridge, Canada, 1907 and 1916

Both design and construction processes deserve special attention with regard to stability design. The failure depicted in Figure 4.4.4 was caused by the temporary removal of a cross frame, performed to install bracing in another location, rather than a design flaw. The removed cross-frame resulted in an excessive unbraced length for the girder during construction (Helwig & Yura, 2015).

Bracing types are numerous. Two of the more frequently encountered in bridge structures are (a) lateral bracing and (b) torsional bracing.

Figure 4.4.5 is a partial framing plan for State Route 26 over Sligo Road and Center Hill Lake in Dekalb County, TN. The cross section for this 4-girder bridge is shown in Figure 4.4.6. Lateral bracing in bays adjacent to the piers was used in this design to reduce the effective span for torsional bracing requirements. The

FIGURE 4.4.3 I-95 over the Brandywine River (Haghani, Al-Emrani, & Heshmati, 2012).

FIGURE 4.4.4   SR-69 over the Tennessee River (Helwig & Yura, 2015).

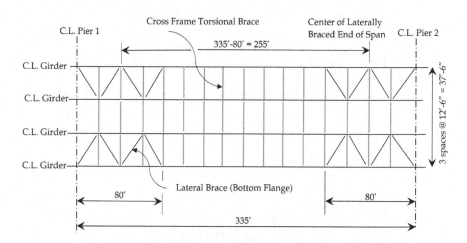

FIGURE 4.4.5   Sligo Bridge partial framing plan.

combination of horizontal laterals near the span ends with cross-frames makes possible this reduced effective span. The support diaphragms are shown in Figure 4.4.6 and a half-section of the intermediate cross-frames is shown in Figure 4.4.7. Four girder lines are being braced by three bays of cross-frames. There are 11 lines of cross-frames between effective supports.

The inverted K-frames depicted in Figure 4.4.6 act as "torsional braces," restraining girder twist, for the welded plate girders. This is a commonly used configuration in steel bridge girders. Design recommendations are not explicit in the AASHTO

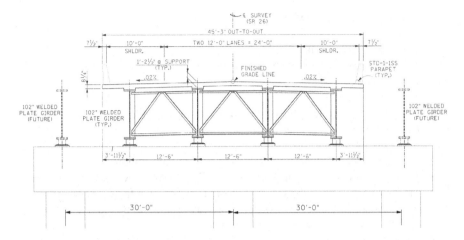

FIGURE 4.4.6 Sligo Bridge cross section – support cross-frames.

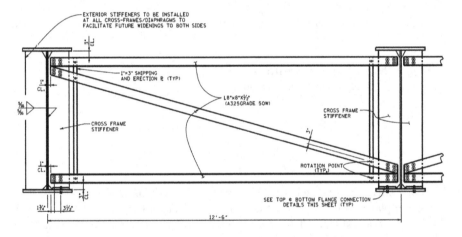

FIGURE 4.4.7 Intermediate Z-frame torsional bracing for the Sligo Bridge.

LRFD Bridge Design Specification, but an excellent guide is provided in the literature (Helwig & Yura, 2015) and is the basis for the material presented here. The guidance is similar to provisions found in AISC 360-16 for steel buildings.

Required torsional bracing stiffness, β_T, and strength, M_{br}, may be determined from Equations 4.4-23 and 4.4-24, respectively.

$$\beta_T = \frac{2.4LM_f^2}{\phi nEI_{eff}C_{bb}^2} \tag{4.4-23}$$

$$M_{br} = \frac{0.005L_bLM_f^2}{nEI_{eff}C_{bb}^2h_o} \tag{4.4-24}$$

In these equations, the following parameters are defined:

- $\phi = 0.75$ (resistance factor)
- M_f is the maximum moment in the longitudinal girder for the span being considered
- L is the span length
- L_b is the unbraced length (spacing of the torsional braces)
- n is the number of torsional braces within the span
- $I_{eff} = I_{yc} + (t/c)I_{yt}$
- I_{yc} is the moment of inertia of the compression flange about a vertical axis
- I_{yt} is the moment of inertia of the tension flange about a vertical axis
- t is the distance from the girder centroid to the tension flange centroid
- c is the distance from the girder centroid to the compression flange centroid
- h_o is the distance between flange centroids $= t + c$
- C_{bb} is a moment modification factor

It is a conservative choice to take $C_{bb} = 1.0$.

The actual system stiffness provided is given by Equation 4.4-25 and is composed of three components:

- β_b is the torsional brace stiffness
- β_{sec} is the web distortional stiffness
- β_g is the in-plane girder system stiffness

$$\frac{1}{\beta_{actual}} = \frac{1}{\beta_b} + \frac{1}{\beta_{sec}} + \frac{1}{\beta_g} \qquad (4.4\text{-}25)$$

The brace stiffness (β_b) for a K-frame torsional brace, web distortional stiffness (β_{sec}), and the in-plane girder system stiffness (β_g) are given by Equations 4.4-26 through 4.4-28. The brace stiffness for other configurations may be found in the handbook (Helwig & Yura, 2015).

$$\beta_b = \frac{2ES^2 h_b^2}{\dfrac{8L_c^3}{A_c} + \dfrac{S^3}{A_h}} \qquad (4.4\text{-}26)$$

$$\beta_{sec} = 3.3 \cdot \frac{E}{h_o}\left(\frac{1.5 h_o t_w^3}{12} + \frac{t_s b_s^3}{12}\right) \qquad (4.4\text{-}27)$$

$$\beta_g = \frac{24(n_g - 1)^2}{n_g} \cdot \frac{S^2 EI_x}{L^3} \qquad (4.4\text{-}28)$$

- A_h is the area of the horizontal member in the K-frame
- A_c is the area of the diagonal member in the K-frame
- L_c is the length of the diagonal member
- S is the girder spacing
- h_b is the height of the K-frame
- E is Young's modulus for steel, 29,000 ksi
- t_w is the girder web thickness
- t_s is the thickness of a full-depth stiffener
- b_s is the width of a one-sided full-depth stiffener of twice the stiffener width if two-sided stiffeners are used
- I_x is the major axis moment of inertia of one girder
- n_g is the number of girders connected by the cross-frames

Example 4.4-4: Stability Bracing for the Sligo Bridge

For the framing plan and cross sections shown in Figures 4.4.5, 4.4.6, and 4.4.7, assess the stiffness and strength requirements for the intermediate cross-frames of the span shown.

The maximum cross-frame spacing used is 25 ft. The actual spacing varies along the 335 ft second span of the five-span bridge (Figure 4.6.3 is an elevation of the bridge showing the span arrangement). The cross-frame members are 2L8 × 6 × ¾ at the Piers and the intermediate cross-frames are Z-type frames (Figure 4.4.7) with chord and diagonal members being L8 × 8 × ½. A structural analysis has shown that the maximum moment in the span for the case considered is 11,424 ft·kips for a single girder, with load factors incorporated. The parameters for this specific example are summarized prior to evaluating the strength and stiffness of the intermediate torsional braces.

- S = 12 ft 6 inches = 150 inches
- L_C = 175 inches
- $A_h = A_c$ = 7.75 in^2
- $t = c$ = 52 inches
- L = 255 ft (effective) = 3,060 inches
- L_b = 25 ft = 300 inches
- n_g = 4
- b_s = 20 inches (double-sided stiffeners)
- t_s = 1.125 inches
- t_w = 0.6875 inches
- h_o = 104 inches
- h_b = 90 inches
- I_x = 320,414 in^4 (average value used in design)
- $I_{yc} = I_{yt}$ = 2,973 in^4
- n = 11
- I_{eff} = 5,946 in^4

For the Z-frame configuration, the handbook (Helwig & Yura, 2015) provides an equation for β_b.

$$\beta_b = \frac{ES^2h_b^2}{\dfrac{2L_c^3}{A_c} + \dfrac{S^3}{A_h}} = \frac{29,000(150)^2\,90^2}{\dfrac{2\times175^3}{7.75} + \dfrac{150^3}{7.75}} = 2,906,301\,\text{in·k/rad}$$

$$\beta_{sec} = 3.3\cdot\frac{29,000}{104}\left(\frac{1.5\times104\times0.6875^3}{12} + \frac{1.125\times20^3}{12}\right) = 694,031\,\text{in·k/rad}$$

$$\beta_g = \frac{24(4-1)^2}{4}\cdot\frac{150^2\times29,000\times320,414}{3,060^3} = 394,023\ \text{in·k/rad}$$

The system stiffness (β_{actual}) may now be determined and compared to the required stiffness (β_T).

$$\frac{1}{\beta_{actual}} = \frac{1}{2,906,301} + \frac{1}{694,031} + \frac{1}{394,023} = \frac{1}{231,328} \rightarrow \beta_{actual} = 231,328\ \text{in·k/rad}$$

$$\beta_T = \frac{2.4(3,060)\left(\dfrac{4}{3}\times11,424\times12\right)^2}{0.75(11)(29,000)(5,946)(1^2)} = 172,477\ \text{in·}\frac{\text{k}}{\text{rad}} < 231,328\ \text{in·}\frac{\text{k}}{\text{rad}}$$

The provided stiffness is adequate. Note that in the equation for required stiffness, the moment was multiplied by 4/3 since the given moment was per girder, and four girders are bring braced by three bays of torsional bracing. This is a conservative approach, and it is not clear if this is the intent of the expressions referenced (Helwig & Yura, 2015). A conservative C_{bb} value equal to 1.00 has been used as well.

$$M_{br} = \frac{0.005(300)(3,060)\left(\dfrac{4}{3}\times11,424\times12\right)^2}{11(29,000)(5,946)(1.0^2)(104)} = 777\ \text{in·kips}$$

This brace moment will produce axial loads in the chords of the Z-frame. The axial forces, one tension and one compression, may be estimated as the moment divided by the Z-frame depth.

$$P_{chord} = \frac{777}{90} = 8.64\ \text{kips}$$

The L8 × 8 × ½ angle is capable of resisting 128 kips in compression and 348 kips in tension, so the strength provided is well above that required. The design is controlled by the stiffness requirements.

4.4.4 FLEXURAL RESISTANCE OF I-GIRDERS

This is not a comprehensive presentation of flexural design requirements, but a general overview to discuss some of the elements of composite construction for steel bridge I-girders. Refer to Section 6.10 of AASHTO LRFD Bridge Design Specifications for a complete presentation of requirements.

Figure 4.4.8 illustrates some of the critical design parameters required for the correct estimation of design flexural strength for composite steel I-girders in positive bending. Positive bending refers to the condition in which top concrete deck is in compression, with the bottom flange of the I-girder in tension. D_c is the depth of the web in compression at the elastic condition, while D_{cp} is the depth of the web in compression at the plastic condition. D_p is the distance from the top of the concrete deck to the plastic neutral axis. D_t is the total depth of the composite girder. The effective flange width of the concrete deck, b_{slab}, is typically taken as equal to the girder spacing for interior girders at uniform spacing, and ½ of the girder spacing plus the overhang for exterior girders. Refer to the AASHTO LRFD Bridge Design Specifications, Section 4.6.2.2, for guidance on situations where this may not be a correct assumption. The concrete in the gap between the top of the girder and the bottom of the deck is often assumed ineffective in section property calculations.

Section properties for I-girders are typically required for four conditions:

a) The steel I-girder acting alone (to carry self-weight, wet concrete, and construction loads)
b) The steel I-girder composite with the deck reinforcing (negative bending)
c) The steel I-girder composite with the concrete deck for long-term loading
d) The steel I-girder composite with the concrete deck for short-term loading

In reinforced concrete design for serviceability requirements, reinforcing steel is transformed into an equivalent area of concrete to determine cracked section

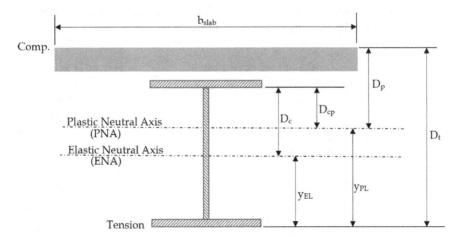

FIGURE 4.4.8 Composite I-girder in positive bending.

properties. The base material is the concrete. The equivalent steel area is the modular ratio, $n = E_S/E_C$, multiplied by the actual steel area. For steel bridge I-girder design, the concrete deck is transformed into an equivalent area of steel to determine composite properties. The base material is the steel. The equivalent concrete width is the actual width divided by n (for short-term property calculations), or by $3n$ (for long-term property calculations). The use of $3n$, rather than n, for long-term properties is a means of accounting for creep and other long-term effects. It is typical to carry loads as follows:

- Self-weight and wet concrete on the I-girder alone
- Sidewalks, parapets, overlays, and utilities dead load on the long-term composite section
- Live loads on the short-term composite section

Flexural requirements for I-girders in AASHTO include (a) cross section proportion limits, (b) constructability requirements, (c) Service Limit State requirements, (d) Fatigue Limit State requirements, and (e) Strength Limit State requirements.

For webs without longitudinal stiffeners, the depth-to-thickness ratio, D/t_w, is limited to no more than 150. For webs with longitudinal stiffeners, the limit is no more than 300. Longitudinal stiffeners are typically not considered for bridges with spans less than about 350 ft or so, though they are permitted.

Both compression and tension flanges are required to satisfy each of Equations 4.4-29 through 4.4-32. I_{yc} is the moment of inertia of the compression flange, and I_{yt} is the moment of inertia of the tension flange, both taken about a vertical axis.

Sections that qualify as "compact" in positive flexure are those satisfying each of Equations 4.4-33 through 4.4-35. F_{yf} is the specified yield strength of the flanges, D is the web depth, and t_w is the web thickness. The nominal flexural resistance of composite sections in positive flexure is given by Equation 4.4-36, and limited in certain situations (See AASHTO Section 6.10.7.1) to that given by Equation 4.4-37. The resistance factor for flexure is 1.00. Both compact and noncompact sections are required to satisfy the ductility requirement of Equation 4.4-38.

$$\frac{b_f}{2t_f} \leq 12.0 \qquad\qquad (4.4\text{-}29)$$

$$b_f \geq D/6 \qquad\qquad (4.4\text{-}30)$$

$$t_f \geq 1.1t_w \qquad\qquad (4.4\text{-}31)$$

$$0.10 \leq \frac{I_{yc}}{I_{yt}} \leq 10 \qquad\qquad (4.4\text{-}32)$$

$$F_{yf} \leq 70 \ ksi \tag{4.4-33}$$

$$\frac{D}{t_w} \leq 150 \tag{4.4-34}$$

$$\frac{2D_{cp}}{t_w} \leq 3.76 \sqrt{\frac{E}{F_{yc}}} \tag{4.4-35}$$

$$M_n = \begin{cases} M_p, \text{if } \dfrac{D_p}{D_t} \leq 0.10 \\ M_p \left(1.07 - 0.7 \dfrac{D_p}{D_t}\right), \text{otherwise} \end{cases} \tag{4.4-36}$$

$$M_n \leq 1.3 R_h M_y \tag{4.4-37}$$

$$\frac{D_p}{D_t} \leq 0.42 \tag{4.4-38}$$

Example 4.4-5: Composite I-Girder Flexural Resistance in Positive Bending

For the composite I-girder shown in Figure 4.4-8, the following parameters are assigned: I, S_{top}, and S_{bott} are the moment of inertia, section modulus at the top of the girder, and section modulus at the bottom of the girder, respectively. The gap between the top of the girder and the bottom of the deck is 3 inches. The area of reinforcement in the concrete deck is 8.91 in², assumed located at mid-thickness of the deck. Young's modulus for the deck is $E_c = 1,820(4)^{0.5} = 3,640$ ksi.

$b_{slab} = 108$ inches $t_{slab} = 8.25$ inches $F_{yf} = 70$ ksi $F_{yw} = 70$ ksi
$D = 75$ inches $f'_c = 4$ ksi $b_{fc} = 24$ inches $t_{fc} = 1.50$ inches
$b_{ft} = 24$ inches $t_{ft} = 2.00$ inches $t_w = 0.75$ inches

- Determine I, S_{top}, and S_{bott} for the girder alone.
- Determine I, S_{top}, and S_{bott} for the girder composite with the rebar.
- Determine I, S_{top}, and S_{bott} for the log-term composite section.
- Determine I, S_{top}, and S_{bott} for the short-term composite section.
- Determine the plastic moment, M_p for positive bending.
- Determine the design flexural resistance, M_n, for positive bending by Equation 4.4-35.
- Check the ductility requirement for positive bending.

For section property calculations, it is convenient to set up a table for the calculations. In the ensuing calculation tables, y_b is the distance from the bottom of the girder to the centroid of the element in question, d is the distance between the cross section centroid and the centroid of the element in question, and I_o is the moment of inertia of a particular element about its centroid axis. The parallel axis theorem is used to determine cross section properties. For the girder alone, after completing columns 2 and 4 in Table 4.4.4, the cross section centroid may be computed:

$$y_{EL} = \frac{5,068.875}{140.25} = 36.14 \text{ inches}$$

So, by the parallel axis theorem:

$$I_x = 26,390 + 122,236 = 148,626 \text{ in}^4$$

$$S_{top} = \frac{148,626}{78.5 - 36.14} = 3,509 \text{ in}^3$$

$$S_{bott} = \frac{148,626}{36.14} = 4,112 \text{ in}^3$$

For the girder composite with the reinforcing in the deck, needed for negative flexure analysis: (Table 4.4.5)

$$y_{EL} = \frac{5,831.554}{149.16} = 39.10 \text{ inches}$$

$$I_x = 148,626 + 20,515 = 169,141 \text{ in}^4$$

$$S_{top} = \frac{169,141}{78.5 - 39.10} = 4,292 \text{ in}^3$$

TABLE 4.4.4

Steel Girder Section Properties – Girder Alone

Element	Area, in²	y_b, inches	Ay_b, in³	d, inches	I_o, in⁴	Ad^2, in⁴
Top flange	36.00	77.75	2,799	41.61	6.75	62,325
Web	56.25	39.50	2,221.875	3.36	26,367	634
Bottom flange	48.00	1.00	48	35.14	16.0	59,277
Sum	140.25		5,068.875		26,390	122,236

TABLE 4.4.5
Steel Girder Section Properties – Girder Composite with Deck Reinforcing

Element	Area, in²	y_b, inches	Ay_b, in³	d, inches	I_o, in⁴	Ad^2, in⁴
Girder	140.25	36.14	5,068.635	2.956	148,626	1,225
Rebar	8.91	85.625	762.919	46.529	0	19,290
Sum	149.16		5,831.554		148,626	20,515

TABLE 4.4.6
Steel Girder Section Properties – Long-Term Composite Section

Element	Area, in²	y_b, inches	Ay_b, in³	d, inches	I_o, in⁴	Ad^2, in⁴
Girder	140.25	36.14	5,068.635	10.36	148,626	15,045
Deck	37.125	85.625	3,178.83	39.13	211	56,837
Sum	177.375		8,247.46		148,837	71,882

$$S_{bott} = \frac{169,141}{39.10} = 4,326 \text{ in}^3$$

For the long-term composite section in positive flexure, use $3n = 3 \times 29,000/3,640$ = 23.9, say $3n = 24$. Although it is permissible to include the reinforcing in positive moment calculations, it is customary practice to reduce the maximum reinforcing in the deck in positive moment sections. The reinforcing bars are neglected in the positive bending, composite section properties in this example. The transformed deck width is 108 inches / 24 = 4.50 inches (Table 4.4.6).

$$y_{EL} = \frac{8,247.46}{177.375} = 46.50 \text{ inches}$$

$$I_x = 148,837 + 71,882 = 220,719 \text{ in}^4$$

$$S_{top} = \frac{220,719}{78.5 - 46.50} = 6,897 \text{ in}^3$$

$$S_{bott} = \frac{220,719}{46.50} = 4,747 \text{ in}^3$$

For the short-term composite section in positive flexure, use $n = 8$. The transformed deck width is 108 inches / 8 = 13.5 inches (Table 4.4.7).

$$y_{EL} = \frac{14,605}{251.625} = 58.04 \text{ in}$$

$$I_x = 149,258 + 152,014 = 301,272 \text{ in}^4$$

$$S_{top} = \frac{301,272}{78.5 - 58.04} = 14,727 \text{ in}^3$$

$$S_{bott} = \frac{301,272}{58.04} = 5,190 \text{ in}^3$$

For the plastic moment, M_p, determine the location of the plastic neutral axis. The general location is first determined by examining the forces in the various yielded elements. As in reinforced concrete design, the stress in the concrete at the Strength Limit State is taken equal to 0.85 times the specified compressive strength.

$$P_{deck} = 0.85 \times 4 \text{ ksi} \times 108 \times 8.25 = 3,029.4 \text{ kips}$$

$$P_{top} = 70 \text{ ksi} \times 24 \times 1.50 = 2,520.0 \text{ kips}$$

$$P_{web} = 70 \text{ ksi} \times 75 \times 0.75 = 3,937.5 \text{ kips}$$

$$P_{bott} = 70 \text{ ksi} \times 24 \times 2 = 3,360 \text{ kips}$$

$$P_{deck} + P_{top} = 3,029.4 + 2,520.0 = 5,549.4 \text{ kips}$$

TABLE 4.4.7
Steel Girder Section Properties – Short-Term Composite Section

Element	Area, in²	y_b, inches	Ay_b, in³	d, inches	I_o, in⁴	Ad^2, in⁴
Girder	140.25	36.14	5,068.635	21.90	148,626	67,285
Deck	111.375	85.625	9,536.48	27.58	632	84,729
Sum	251.625		14,605		149,258	152,014

$$P_{web} + P_{bott} = 3,937.5 + 3.360 = 7,297.5 \text{ kips}$$

Since $P_{deck} + P_{top} < P_{web} + P_{bott}$, then part of the web at the top must be in compression in order to balance the internal forces. Let x be the distance from the top of the web to the plastic neutral axis. Thus, x is the depth of the web in compression at the plastic condition. Sum forces above and below x to determine the exact location.

$$3,029.4 + 2,520.0 + 3,937.5 \cdot \frac{x}{75} = 3,937.5 \cdot \frac{75 - x}{75} + 3,360$$

$$D_{cp} = x = 16.65 \text{ inches}$$

With the plastic neutral axis located, sum moments above and below the PNA to determine M_P.

$$M_P = 3,029.4 \left(25.27 \text{ in}\right) + 2,520.0 \left(17.40 \text{ in}\right) + 3,937.5 \cdot \frac{16.65}{75} \left(\frac{16.65}{2} \right)$$

$$+ 3,937.5 \cdot \frac{58.35}{75} \left(\frac{58.35}{2} \right) + 3,360 \left(59.35\right)$$

$$= 416,473 \text{ in} \cdot \text{kips} = 34,706 \text{ ft} \cdot \text{kips}$$

The distance from the top of the deck to the PNA is $D_p = 8.25+3+1.5+16.65 = 29.40$ inches. The total composite depth is $D_t = 2+75+1.5+3+8.25 = 89.75$ inches. So the ductility requirement is assessed as follows:

$$\frac{D_{cp}}{D_t} = \frac{29.40}{89.75} = 0.328 < 0.42 \rightarrow OK$$

$$M_n = 34,706 \left(1.07 - 0.7 \times 0.328\right) = 29,177 \text{ ft} \cdot \text{kips}$$

The example presented here is far from comprehensive but serves to introduce the concept of flexural resistance and typical calculations for composite steel bridge I-girders. While I-girders are employed more frequently, box (tub) steel girders are found in practice as well.

4.4.5   SHEAR RESISTANCE OF STEEL I-GIRDERS

For equations and issues related to shear resistance of I-girders, refer to the previous section on field splice design as well as to Section 6.10.9 of the AASHTO LRFD Bridge Design Specifications.

4.5 REINFORCED CONCRETE SUBSTRUCTURES

Reinforced concrete substructures are frequently used for most bridge types. Hammerhead piers and multi-post piers both are popular. A full appreciation of the material presented here requires that the reader be familiar with basic concepts of reinforced concrete design. Of particular interest is flexure-shear interaction in deep concrete members. While the strut-and-tie method is a relatively recent development and useful for deep members, the focus here will be on section analysis by Response 2000 using the Modified Compression Field Theory (MCFT) (Bentz, 2000).

Example 4.5-1: Bent Cap Analysis – State Route 52 Bridge

Figure 4.5.1 is an excerpt from the construction plans for the bent cap design for State Route 52 over Branch in Clay County, TN. Figure 4.5.2 shows the cross section of the bent cap as modeled in Response 2000. The factored loads at the column face, 14 ft 9 inches from the end of the cantilever cap, are: $M_u = 12,664$ ft-kips and $V_u = 1,585$ kips.

The moment to shear ratio is an important parameter in the theoretical calculation of resistance. The ratio for the controlling load combination for the cap is $M/V = 7.99$ ft. The Response 2000 analysis reports a nominal flexural resistance at this M/V ratio of $M_n = 15,752$ ft·kips with a coincident nominal shear resistance of $V_n = 1,971$ kips. If the effect of shear on flexural resistance were ignored, Response 2000 reports a nominal flexural resistance of $M_n = 20,457$ ft·kips. It becomes clear

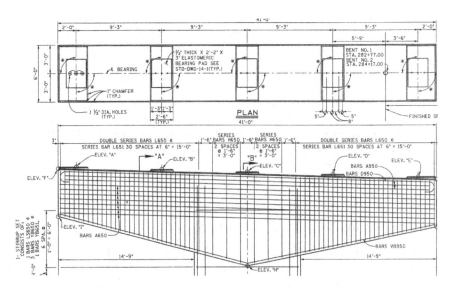

FIGURE 4.5.1　Bent cap for State Route 52 over Branch in Clay County, TN.

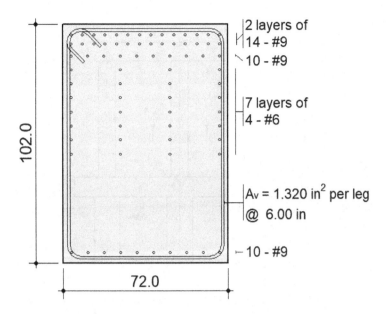

Labels on figure:
2 layers of
14 - #9
10 - #9

7 layers of
4 - #6

Av = 1.320 in^2 per leg
@ 6.00 in

10 - #9

102.0

72.0

FIGURE 4.5.2 Response 2000 Bent Cap cross section model.

that neglecting the effect of shear on the design resistance of such concrete elements can result in serious errors.

Note that the stirrups shown in Figure 4.5.2 are the equivalent of six legs of No. 6 bars spaced at 6 inches and that the default stress-strain relationship for reinforcement in Response 2000 incorporates strain hardening. The AASHTO LRFD Bridge Design Specifications do not permit stress beyond yield in determining nominal capacities. So the default stress-strain relationship in Response 2000 was first modified to remove strain-hardening effects in order to reach the above conclusions regarding nominal resistance.

Example 4.5-2: Sligo Bridge Concrete Strut and Shaft Analyses

Figure 4.5.3 is an elevation of the Pier used for State Route 26 over Sligo Road and Center Hill Lake in Dekalb County, TN. A cross section of the 8-ft deep strut between the columns and the shafts is shown in Figure 4.5.4. During strong ground shaking from seismic loading, the strut loads are large, with a moment to shear ratio, $M/V = 16.6$ ft. Response 2000 provides a nominal flexural resistance, $M_n = 23,098$ ft·kips with flexure-shear interaction included. Neglecting the interaction results in a predicted nominal flexural resistance, $M_n = 27,725$ ft·kips. Once again, the error in neglecting the interaction between shear and flexure is significant.

Figure 4.5.5 depicts a cross section of the shafts in the water for the Sligo Bridge. Grade 75 bars were used for longitudinal reinforcement in the shafts. Large axial forces, combined with shear and flexure, control the design of the shafts. In this case, the moment to shear ratio, $M/V = 209$ ft, is much larger than the value for the strut. With an axial compression of 3,612 kips, Response 2000 results for

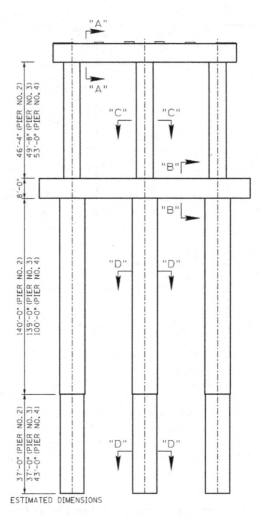

FIGURE 4.5.3   State Route 26 over Sligo Road and Center Hill Lake – pier elevation.

nominal flexural resistance are M_n = 44,889 ft·kips incorporating shear-flexure interaction, and M_n = 46,688 ft·kips, ignoring shear-flexure interaction. The difference in results for this element with a large M/V ratio, while much less than the difference for the strut, should be addressed in design. It would be wise to incorporate the interaction among axial force, shear, and moment in most reinforced concrete element subjected to all three, given the capabilities of modern software.

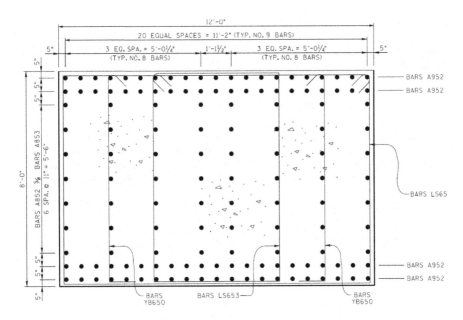

FIGURE 4.5.4   Sligo Bridge Strut cross section.

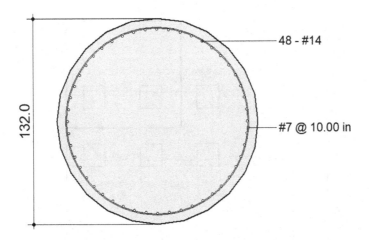

FIGURE 4.5.5   Sligo Bridge drilled shaft cross section.

4.6 FOUNDATION SYSTEMS

Commonly used foundation systems for bridges include point-bearing piles with cap, friction piles with cap, spread footings on rock, drilled shafts, and pile bents.

Example 4.6-1: Pile Cap Foundation

Figure 4.6.1 details dimensions for an example pile cap and pile arrangement. Square prestressed concrete piles are indicated. Suppose that the loads on the column, which is denoted by the dotted lines in the figure, are as follows. The factored self-weight of the pile cap is already incorporated into the vertical load, P_u.

- P_u = 1,500 kips
- M_{ux} = 1,250 ft·kips
- M_{uy} = 750 ft·kips

The loads have already been factored to the Strength Limit State. The task required is to estimate the maximum and minimum axial pile loads.

$N = 16$ piles

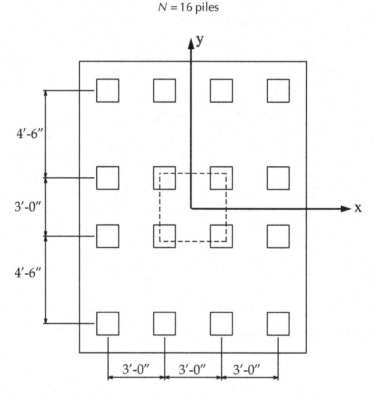

FIGURE 4.6.1 Pile cap example.

$$I_x = 8\left(1.5^2 + 6.0^2\right) = 306 \text{ pile} \cdot \text{ft}^2$$

$$I_y = 8\left(1.5^2 + 4.5^2\right) = 180 \text{ pile} \cdot \text{ft}^2$$

$$P_{pile} = \frac{1,500}{16} \pm \frac{1,250(6)}{306} \pm \frac{750(4.5)}{180} = 94 \pm 25 \pm 19$$

$$P_{Max} = 138 \text{ kips, compression}$$

$$P_{Min} = 50 \text{ kips, compression}$$

Should similar calculations result in a negative value for the minimum pile load, the pile is in tension.

Drilled shafts are often appropriate for bridge substructures. For rock with many cavities and for substructures in deep water, cased drilled shafts may be ideal.

Example 4.6-2: Drilled Shaft Foundation

A partial set of AASHTO, 8th edition, requirements for geotechnical design of drilled shafts is summarized in Equations 4.6-1 and 4.6-2. With intact rock at least 2B below the shaft tip under normal conditions, full tip resistance (q_p) and side resistance (q_s) for shafts socketed into rock are given by the following equations:

$$q_p = 2.5q_u \tag{4.6-1}$$

$$q_s = p_a\sqrt{\frac{q_u}{p_a}} \tag{4.6-2}$$

q_u is the uniaxial compressive strength of the rock.
q_a is the atmospheric pressure, taken as 2.12 ksf.

The top 5 ft of socket is to be disregarded in side resistance calculations. The Drilled shaft for Pier 2 on SR-26 over Sligo Road and Center Hill Lake is shown in Figure 4.6.2. The estimated elevations are 527 (top of lakebed), 521 (top of weathered rock), 511 (top of competent rock), and 481 (shaft tip). The Strength Limit State resistance factor, ϕ, is 0.45 for both tip and side resistance. The Strength Limit State axial load, with two future girders installed, is 7,747 kips. Figure 4.6.3 depicts an elevation of the bridge. The socket diameter = 10.5 ft. End spans are 270-ft and interior spans are 335-ft for a total bridge length of 1,545-ft.

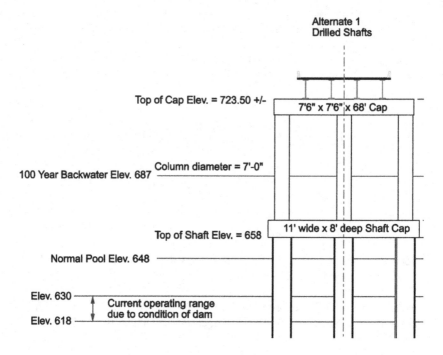

Alternate 1 Drilled Shafts

Top of Cap Elev. = 723.50 +/-

7'6" x 7'6" x 68' Cap

Column diameter = 7'-0"

100 Year Backwater Elev. 687

Top of Shaft Elev. = 658

11' wide x 8' deep Shaft Cap

Normal Pool Elev. 648

Elev. 630
Elev. 618

Current operating range due to condition of dam

FIGURE 4.6.2 Pier 2 elevation for State Route 26 over Sligo Road and Center Hill Lake.

FIGURE 4.6.3 Elevation of State Route 26 over Sligo Road and Center Hill Lake.

Nominal tip resistance:

$$A_{tip} = \frac{\pi \left(10.5\right)^2}{4} = 86.59 \text{ ft}^2$$

$$q_p = 2.5 \times 46 \text{ tsf} \times \frac{2k}{\text{ton}} = 230 \text{ ksf}$$

$$P_p = 86.59 \times 230 = 19,916 \text{ kips}$$

Nominal side resistance:

   Top of weathered rock = 521 ft
   Socket tip = 481 ft
   Total socket length = 521 − 481 = 40 ft

Neglect the top 5 ft and use L_{socket} = 35 ft

$$A_{side} = \pi (10.5)(35) = 1,154 \text{ ft}^2$$

$$q_s = 2.12 \sqrt{\frac{46 \times 2}{2.12}} = 13.97 \text{ ksf}$$

$$P_s = 1,154 \times 13.97 = 16,124 \text{ kips}$$

(a)
$$\phi P_n = 0.45 \big[1.00 (19,916) + 0.00 (16,124) \big]$$
$$= 8,962 \text{ kips} > P_u = 7,747 \text{ kips} \cdot \text{OK}$$

(b)
$$\phi P_n = 0.45 \big[0.50 (19,916) + 0.50 (16,124) \big]$$
$$= 8,109 \text{ kips} > P_u = 7,747 \text{ kips} \cdot \text{OK}$$

(c)
$$\phi P_n = 0.45 \big[0.00 (19,916) + 1.00 (16,124) \big]$$
$$= 7,256 \text{ kips} < P_u = 7,747 \text{ kips} \cdot \text{NG}$$

It is not possible to take 100% of the load in side resistance. Therefore, take measures to clean out and inspect the shaft bottom prior to setting that shaft cage and pouring the shaft concrete. For this example, three cases were considered: 100% of the tip resistance, 50% each of tip and side resistance, and 100% of side resistance.

Figure 4.6.4 is a photograph of the bridge upon which this example was based.

For driven friction piles, AASHTO (AASHTO, LRFD Bridge Design Specifications, 2012) provides multiple methods for estimating design resistance. Two methods for estimating geotechnical resistance during driving included in the design specifications are the Gates Method (Equation 4.6-3) and the Engineering News Record (ENR) Method (Equation 4.6-4). The resistance factor at the Strength Limit State for the Gates Method is 0.40, while that for the ENR Method is 0.10.

FIGURE 4.6.4 State Route 26 over Center Hill Lake in Dekalb County, TN (TDOT).

$$R_{ndr} = 1.75\sqrt{E_d}\,\log_{10}\left(10N_b\right) - 100, Gates \qquad (4.6\text{-}3)$$

$$R_{ndr} = \frac{12E_d}{s+0.1},\ ENR \qquad (4.6\text{-}4)$$

The parameters in the Gates Method, Equation 4.6-3, are as follows:

- R_{ndr} = nominal pile resistance, kips
- E_d = hammer energy (ram weight times stroke height), ft·lb
- N_b = number of blows for 1 inch of pile set

The parameters in the ENR Method, Equation 4.6-4, are as follows:

- R_{ndr} = nominal pile resistance, kips
- E_d = hammer energy (ram weight times stroke height), ft·kips
- s = pile set, inches/blow

Note, in particular, the difference in units for E_d in the two methods.

Example 4.6-3: Pile Resistance from Driving Logs

Suppose a 5.51-kip ram with 12.04-ft stroke is used to drive a pile and a blow count of 24 blows/ft is indicated at the end of the driving. Estimate the Strength Limit State pile resistance using the Gates Method and the ENR Method.
Gates Method:

$$E_d = 5,510 \text{ lbs} \times 12.04 \text{ ft} = 66,340 \text{ ft} \cdot \text{lb}$$

$$N_b = 24 \text{ blows/ft} \div 12 = 2.0 \text{ blows/in}$$

$$\phi R_{ndr} = 0.40 \left[1.75 \sqrt{66,340} \; \log_{10} \left(10 \times 2.00 \right) - 100 \right] = 195 \text{ kips} = 97 \text{ tons}$$

ENR Method:

$$E_d = 5.51 \text{ kips} \times 12.04 \text{ ft} = 66.34 \text{ ft} \cdot \text{kips}$$

$$s = \frac{1}{N_b} = \frac{1}{2.00} = 0.50 \text{ in/blow}$$

$$\phi R_{ndr} = 0.10 \left[\frac{12 \times 66.34}{0.50 + 0.10} \right] = 133 \text{ kips} = 66 \text{ tons}$$

For variations on the formulas for various driving equipment specifications, refer to the AASHTO Specifications and manufacturer's literature.

4.7   SEISMIC ANALYSIS AND DESIGN OF BRIDGES – PUSHOVER ANALYSIS

Bridge design for seismic effects has increasingly become displacement-based, as opposed to original, decades-old force-based provisions. Pushover analysis by computer modeling and approximate hand calculations are both useful.

Approximate pushover analysis of concrete piers by hand may be accomplished using an analysis incorporating the Mander model for confined concrete and procedures outlined in the literature (Priestley, Calvi, & Kowalsky, 2007) and reproduced here in Equations 4.7-1 through 4.7-17.

$$\phi_y \cong \begin{cases} 2.25\left(\varepsilon_y \big/ D\right), \text{ circular columns} \\ 2.10\left(\varepsilon_y \big/ D\right), \text{ rectangular columns} \end{cases} \tag{4.7-1}$$

$$\frac{c}{D} \cong 0.20 + 0.65\frac{P}{f'_{ce}A_g} \tag{4.7-2}$$

$$\varepsilon_{cu} = \begin{cases} 0.004 + \dfrac{1.4\rho_v f_{yh}\varepsilon_{su}}{f'_{cc}}, \text{ damage control limit state} \\ 0.004, \text{ serviceability limit state} \end{cases} \tag{4.7-3}$$

$$\varepsilon_{su} = \begin{cases} 0.06 \text{ to } 0.09, \text{ damage control limit state} \\ 0.015, \text{ serviceability limit state} \end{cases} \tag{4.7-4}$$

$$\rho_{cc} = \frac{A_s}{A_c} = \frac{\text{Area of longitudinal reinforcement}}{\text{Area of core enclosed by centerlines of hoop or spiral}} \tag{4.7-5}$$

$$\rho_v = \begin{cases} \dfrac{A_{sp}\pi d_s}{\frac{\pi}{4}d_s^2 s} = \dfrac{4A_{sp}}{d_s s} \to \text{circular hoops or spirals} \\ \rho_{vx} + \rho_{vy} = \dfrac{A_{sx}}{sh_{cy}} + \dfrac{A_{sy}}{sh_{cx}} \to \text{rectangular hoops} \end{cases} \tag{4.7-6}$$

$$f_{cc} = f'_{co}\left(-1.254 + 2.254\sqrt{1 + \frac{7.94f'_l}{f'_{co}}} - 2\frac{f'_l}{f_{co}}\right) \tag{4.7-7}$$

$$k_e = \begin{cases} \dfrac{\left(1 - \dfrac{s'}{2d_s}\right)^2}{1 - \rho_{cc}} \to \text{circular hoop effectiveness coefficient} \\ \dfrac{1 - \dfrac{s'}{2d_s}}{1 - \rho_{cc}} \to \text{circular spiral effectiveness coefficient} \\ \dfrac{\left(1 - \sum\dfrac{(w_i')^2}{6b_c d_c}\right)\left(1 - \dfrac{s'}{2b_c}\right)\left(1 - \dfrac{s'}{2d_c}\right)}{1 - \rho_{cc}} \to \text{rectangular hoop coefficient} \end{cases} \tag{4.7-8}$$

$$f_l = \frac{1}{2}k_e\rho_v f_{yh} \to \text{lateral confining stress on concrete} \tag{4.7-9}$$

$$\Delta_y = \frac{1}{3}\phi_y\left(L_c + L_{SP}\right)^2 \tag{4.7-10}$$

$$\Delta_P = \left(\phi_u - \phi_y\right)L_P\left(L_c + L_{SP} - \frac{L_P}{2}\right) \tag{4.7-11}$$

$$k = 0.20\left(\frac{f_u}{f_y} - 1\right) \le 0.08 \tag{4.7-12}$$

$$L_P = kL_c + L_{SP} \ge 2L_{SP} \tag{4.7-13}$$

$$L_{SP} = 0.15f_{ye}d_{bl} \tag{4.7-14}$$

$$\Delta_u = \Delta_y + \Delta_P \tag{4.7-15}$$

$$\mu = \frac{\Delta_u}{\Delta_y} \tag{4.7-16}$$

$$\phi_u = Min\begin{cases} \dfrac{\varepsilon_{cu}}{c} \\ \dfrac{\varepsilon_{su}}{d-c} \end{cases} \tag{4.7-17}$$

- L_c is the distance from critical section to point of contraflexure
- L_{SP} is the strain penetration distance
- L_P is the plastic hinge length
- d is the column depth
- c is the distance from the compression face of the column to the neutral axis
- d_s is the hoop or spiral diameter measured to the center of the hoop or spiral
- s is the hoop or spiral pitch measure to the center of the spiral or hoop
- s' is the clear hoop or spiral pitch
- A_{sp} is the area of the spiral bar
- f'_{co} is the specified concrete strength at 28 days
- b_c is the out-to-out width measured to the center of the rectangular hoop
- d_c is the out-to-out height measured to the center of the rectangular hoop
- w' is the clear spacing between adjacent longitudinal bars

Example 4.7-1: Pushover Analysis of a Bridge Pier

Figure 4.7.1 depicts a bridge pier for which both computer and hand solutions will be completed and compared. It is required to estimate the bilinear force displacement curve for the pier and to estimate the plastic shear, V_P. Data for the pier are as follows:

- L_C = 655 cm (258 inches)
- H_{cg} = 1707 cm (672 inches)
- S = 514 cm (202 inches = 16.875 ft)
- D = 137 cm (54 inches) column diameter
- Each column has 23 32 mm (#10) longitudinal bars
- Transverse bars are 19 mm spirals spaced at 15.2 cm pitch (#6 at 6 inches)
- f'_{ce} = 35.8 MPa (5.2 ksi) expected concrete strength
- f_{ye} = 469 MPa (68 ksi) expected rebar yield stress
- Initial column loads are 1,088 kips for each exterior column and 1,450 kips for the interior column

Determine the yield curvature of the columns.

$$\phi_y \cong \frac{2.25}{54} \cdot \frac{68}{29,000} = 0.0000977 \text{ in}^{-1}$$

Determine the volumetric ratio of the transverse spirals.

$$d_s = 54 - (2 \times 2.5)(\text{clear}) - 0.75 (\text{spiral diameter}) = 48.25 \text{ inches}$$

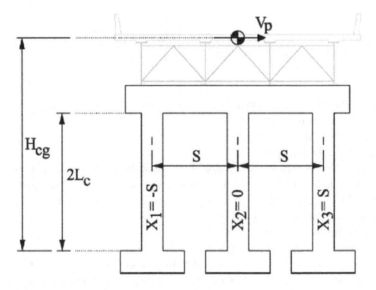

FIGURE 4.7.1   Pushover analysis example pier.

$$\rho_v = \frac{4 \cdot 0.44}{48.25 \cdot 6} = 0.00608$$

Determine the longitudinal reinforcement ratio.

$$A_c = \frac{\pi \cdot 48.25^2}{4} = 1,828 \text{ in}^2$$

$$\rho_{cc} = \frac{23 \cdot 1.27}{1,828} = 0.01598$$

Determine the spiral confinement effectiveness coefficient.

$$s' = 6'' - 0.75'' = 5.25''$$

$$k_e = \frac{1 - \dfrac{5.25}{2 \times 48.25}}{1 - 0.01598} = 0.961$$

Determine the lateral confining stress. Use the minimum specified yield stress (60 ksi) rather than the expected yield stress (68 ksi) for this calculation to be conservative.

$$f'_l = \frac{1}{2} \times 0.961 \times 60 \times 0.00608 = 0.175 \text{ ksi}$$

Determine the confined concrete strength.

$$f_{cc} = 5.2\left(-1.254 + 2.254\sqrt{1 + \frac{7.94 \times 0.175}{5.2}} - 2\frac{0.175}{5.2}\right) = 6.33 \text{ ksi}$$

Determine the ultimate usable concrete strain for the damage limit state.

$$\varepsilon_{cu} = 0.004 + \frac{1.4(0.00608)(60)(0.06)}{6.33} = 0.00884 \text{ in/in}$$

Determine the average column axial load and the depth to the neutral axis.

$$P = \frac{1}{3}(2 \times 1,088 + 1,450) = 1,209 \text{ kips}$$

$$A_g = \frac{\pi \cdot 54^2}{4} = 2,290 \text{ in}^2$$

$$\frac{c}{D} \cong 0.20 + 0.65\frac{1,209}{5.2 \cdot 2,290} = 0.266 \rightarrow c = 0.266 \times 54 = 14.4 \text{ inches}$$

Determine the steel-controlled and the concrete-controlled ultimate curvatures.

$$d = 54 \text{ inches} - 2.5 \text{ inches} - 0.75 \text{ inch} - 1.27 \text{ inches}/2 = 50.1 \text{ inches}$$

$$\phi_u = Min \begin{cases} \dfrac{0.00884}{14.4} \\ \dfrac{0.06}{50.1 - 14.4} \end{cases} = 0.000614 \text{ in}^{-1}$$

Determine the analytical plastic hinge length and the strain penetration length.

$$L_P = 0.08 \times 258 + 0.15 \times 68 \times 1.27 = 20.64 + 12.95 = 33.6 \text{ inches}$$

$$L_{SP} = 12.95 \text{ inches}$$

Determine the yield and ultimate displacements. Equations 4.7-10 and 4.7-11 estimate yield and plastic displacements between points of zero and maximum moment. Assume a point of contraflexure at mid-height of the columns, resulting in equal displacements between (a) column bottom and mid-height and (b) mid-height and column top, each equal to values given by Equations 4.7-10 and 4.7-11.

$$\Delta_y = 2\left[\frac{1}{3} \times 0.0000977\left(258 + 12.95\right)^2\right] = 4.78 \text{ inches}$$

$$\Delta_P = 2\left[\left(0.000614 - 0.0000977\right)\left(33.6\right)\left(258 + 12.95 - 33.6/2\right)\right] = 8.81 \text{ inches}$$

$$\Delta_u = 4.78 + 9.34 = 13.59 \text{ inches}$$

$$\mu = \frac{13.59}{4.78} = 2.84$$

For the plastic shear, it is necessary to have a relationship between moment resistance, M_n, and axial load, P, in the columns. A section analysis program is useful for this. For the initial column loads, the flexural resistance values, using expected material strengths (1.3 times the specified concrete compressive strength and 68 ksi for the steel yield stress) are as follows:

- $P = 1,088$ kips, $M_n = 5,045$ ft·kips, exterior columns
- $P = 1,450$ kips, $M_n = 5,339$ ft·kips, interior column

The distance from the center of gravity to mid-height of the columns, the assumed point of contraflexure, is $H_{cg} - L_C = 672$ inches $- 258$ inches $= 414$ inches $= 34.5$ ft. The estimated shear in each column is simply the moment resistance divided by L_C. For the first iteration, the column shears are as follows:

$$V_1 = V_3 = \frac{5,045 \text{ ft-k}}{21.5'} = 235 \text{ kips}$$

$$V_2 = \frac{5,339 \text{ ft-k}}{21.5'} = 248 \text{ kips}$$

For iteration number 1, the estimated plastic shear is $2 \times 235 + 248 = 718$ kips. The plastic shear creates an overturning effect, which increases the axial load on column 3 and decreases the axial load on column 1 by δP. The axial load in column 2, at the center of the pier, remains unchanged.

$$\delta P = \frac{718 \times 34.5}{2 \times 16.875} = 734 \text{ kips}$$

The new column loads, the corresponding flexural resistances, and the new column shears may now be determined.

$$P_1 = 1,088 - 734 = 354 \text{ kips} \rightarrow M_n = 4,305 \text{ ft} \cdot \text{k}$$

$$P_2 = 1,450 \text{ kips} \rightarrow M_n = 5,330 \text{ ft} \cdot \text{k}$$

$$P_3 = 1,088 + 734 = 1,822 \text{ kips} \rightarrow M_n = 5,620 \text{ ft} \cdot \text{k}$$

$$V_1 = \frac{4,305}{21.5} = 200 \text{ kips}$$

$$V_2 = \frac{5,330}{21.5} = 248 \text{ kips}$$

$$V_3 = \frac{5,620}{21.5} = 261 \text{ kips}$$

$$V_P = \sum V_i = 200 + 248 + 261 = 709 \text{ kips}$$

A third iteration produces less than a 1% difference in the estimated plastic shear, and further iteration is not warranted. Each column should be checked for design

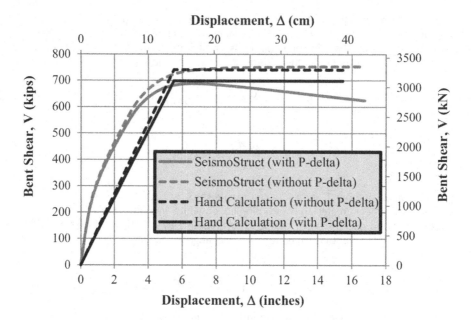

FIGURE 4.7.2   Hand calculation pushover versus SeismoStruct pushover.

shear resistance at the respective axial load levels. For these relatively tall pier columns, the design shear resistance would be found to be much larger than the shear demands indicated in iteration number 2. The plastic shears for design would need to be increased by an overstrength factor, usually taken between 1.2 (for A 706 reinforcing bars) or 1.4 (for A 615 reinforcing bars), depending on the type of reinforcing steel used (AASHTO, 2011).

Figure 4.7.2 depicts a hand-calculated solution compared to a computer solution using SeismoStruct (Seismosoft, 2020) nonlinear analysis with force-based inelastic elements for the columns.

4.8   SEISMIC ISOLATION OF BRIDGES

Seismic isolation devices commonly used for bridges include lead-rubber bearings and friction-pendulum bearing devices.

A strategy that has been shown to hold promise is that of partial isolation. Isolation devices located at intermediate piers, with abutments remaining fixed, produces a bridge structure for which substantial savings in pier costs may be realized.

Figure 4.8.1 shows a pile bent substructure. For such bridges, partial isolation has been shown to be an effective strategy under favorable conditions (Huff, 2016).

A bridge on Interstate 40 near Jackson, TN, has been constructed using partial isolation and is the subject of a paper at the 2017 International Bridge Conference (Huff & Shoulders, 2017). Oversized piers for architectural consideration resulted in extremely large plastic shears, for which piling would need to be designed. Partial

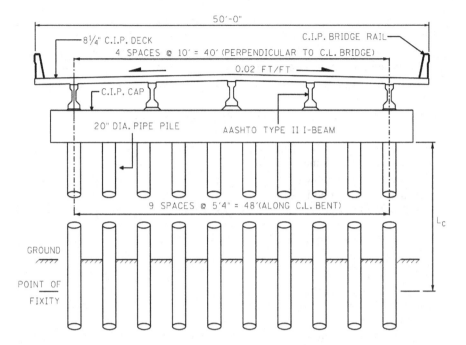

FIGURE 4.8.1 Pile bent substructure.

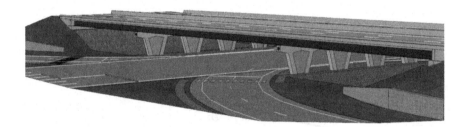

FIGURE 4.8.2 Interstate 40 over State Route 5 in Jackson, TN

isolation with lead-rubber bearings permitted a dramatic reduction in piling costs at the intermediate piers of the three-span bridge shown in Figure 4.8.2.

Approximate, hand calculation methods were used for preliminary design. Final design was confirmed using nonlinear response history analysis with a set of 14 ground motion pairs scaled to achieve a level of shaking consistent with that represented by the response spectrum corresponding to a 2,500-year mean recurrence interval.

The Hernando de Soto Bridge carrying Interstate 40 over the Mississippi River near Memphis, TN, has been seismically retrofitted with friction pendulum isolators at three piers supporting a steel-tied arch and lead-rubber bearings at approach spans.

Figure 4.8.3 depicts a typical load-deformation curve for lead-rubber bearing (LRB) and friction-pendulum system (FPS) isolation bearings. Lead-rubber bearings are constructed of elastomeric material with a central lead plug, which effectively adds damping to the system when yielded in shear.

For LRB isolators, the pertinent parameters defining the behavior are as follows:

f: factor to account for post-yield stiffness of the lead core (1.1, typical)
G: elastomer shear modulus (50–300 psi, typical)
G_p: shear modulus of lead plug (21.75 ksi, typical)
T_r: total rubber (elastomer) thickness
A_b: bonded area of rubber
d_b: diameter of circular bearing
d_L: diameter of lead plug ($d_b/6$ to $d_b/3$, typically)
f_{yL}: yield stress in shear for lead (1.3–1.5 ksi, typical)
ψ: stress modifier for lead plug
   1 for EQ-load
   2 for wind/braking
   3 for thermal expansion
α: post-yield stiffness ratio, typical values:
   0.10 for EQ
   0.125 for wind/braking
   0.20 for thermal expansion
γ_c: shear strain due to compression

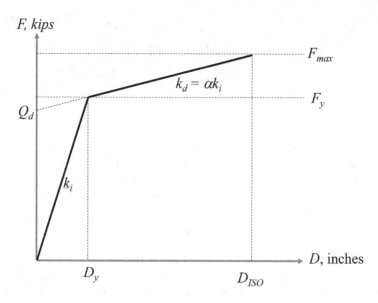

FIGURE 4.8.3   Bilinear isolator load-deformation curve.

γ_r: shear strain due to rotation
$\gamma_{s,s}$: shear strain due to nonseismic displacement
$\gamma_{s,eq}$: shear strain due to seismic displacement
t_i: thickness of an individual elastomer layer
θ: bearing rotation (include a 0.005 radian contingency)
B: bonded plan dimension in the direction of loading (d_b for circular)
S: shape factor; bonded plan area divided by the side area free
to bulge
D_r: shear strain factor
   0.375, circular bearing
   0.500, rectangular bearing
D_C: shape coefficient
   1.000, circular bearing
   1.000, rectangular bearing
σ_s: compressive stress $= P/A_b$

Design criteria for LRB isolators used in bridge structures may be found in AASHTO design specifications (AASHTO, 2014) supplemented with literature on the subject (Buckle, Constantinou, Dicleli, & Ghasemi, 2006). A subset of applicable criteria are summarized in Equations 4.8-1 through 4.8-17. Loads that contribute to the "static" components of deformation include wind, dead load, and thermal effects. Loads that are assumed to contribute to the "cyclic" components of deformation include live load, braking forces, and seismic effects.

$$A_b = \frac{\pi\left(d_b^2 - d_L^2\right)}{4} \tag{4.8-1}$$

$$k_d = f\frac{GA_b}{T_r} \tag{4.8-2}$$

$$k_i = \frac{G_p A_p + GA_b}{T_r} \tag{4.8-3}$$

$$F_y = \frac{1}{\psi} \cdot f_{yL} \cdot \frac{\pi d_L^2}{4} \tag{4.8-4}$$

$$S = \frac{d_b^2 - d_L^2}{4 d_b t_i} \tag{4.8-5}$$

$$\gamma_c = \frac{D_c \sigma_s}{GS} \tag{4.8-6}$$

$$\gamma_r = \frac{D_r B^2 \theta}{t_i T_r} \tag{4.8-7}$$

$$\gamma_{s,s} = \frac{\Delta_s}{T_r} \tag{4.8-8}$$

$$\gamma_{s,eq} = \frac{D_{ISO}}{T_r} \tag{4.8-9}$$

$$\left(\gamma_c + \gamma_r + \gamma_{s,s}\right)_{static} + 1.75\left(\gamma_c + \gamma_r + \gamma_{s,s}\right)_{cyclic} \le 5.0 \tag{4.8-10}$$

$$\left(\gamma_c\right)_{static} \le 3.0 \tag{4.8-11}$$

$$\left(\gamma_c + 0.50\ \gamma_r + \gamma_{s,s} + \gamma_{s,\ eq}\right)_{total} \le 5.5 \tag{4.8-12}$$

$$K_{EFF} = k_d + \frac{Q_d}{D_{ISO}} \tag{4.8-13}$$

$$\xi_{EFF} = \frac{2Q_d\left(D_{ISO} - D_y\right)}{\pi\left(D_{ISO}\right)^2 K_{EFF}} \tag{4.8-14}$$

$$D_y = \frac{Q_d}{k_d} \cdot \frac{\alpha}{1-\alpha} \tag{4.8-15}$$

$$F_y = \frac{Q_d}{1-\alpha} \tag{4.8-16}$$

$$B_L = \left(\frac{\xi_{EFF}}{0.05}\right)^{0.30} \tag{4.8-17}$$

Example 4.8-1: Lead-Rubber Bearing Isolator Design

The properties of the bearings used for Interstate 40 over State Route 5 in Madison County, TN (Figure 4.8.2), are given below. Refer to Figure 4.8.4 for isolator bearing dimensions. Check the bearings for AASHTO criteria for isolation bearings.

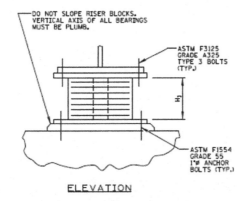

DO NOT SLOPE RISER BLOCKS.
VERTICAL AXIS OF ALL BEARINGS
MUST BE PLUMB.

ASTM F3125
GRADE A325
TYPE 3 BOLTS
(TYP.)

H_1

ASTM F1554
GRADE 55
1"ø ANCHOR
BOLTS (TYP.)

ELEVATION

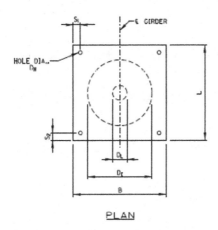

S_L
¢ GIRDER
HOLE DIA.
D_N
L
S_B
D_L
D_f
B

PLAN

FIGURE 4.8.4 LRB isolator for I40 over SR5 in Jackson, TN.

- d_b = 22.5 inches
- d_l = 3.75 inches
- H_1 = 8.75 inches (nine ¾-inch internal elastomer layers, two 3/8-inch cover layers, ten 1/8-inch steel layers)
- Δ_S = 0.75 inches (non-seismic lateral deformation demand)
- D_{ISO} = 8.30 inches (seismic isolator deformation demand from structural analysis)
- θ = 0.007 rad (dead + live + miscellaneous rotation)
- P_{DL} = 258 kips, P_{LL} = 40 kips, P_{EQ} = 11 kips
- f_{yL} = 1,450 psi, lead plug shear yield stress
- G_p = 21,750 psi, lead plug shear modulus
- G = 100 psi, specified elastomer shear modulus

For this structure, partial isolation resulted in a reduction in seismic shear transmitted to each pier from 8,044 kips to 1,175 kips per pier.

$$A_b = \frac{\pi\left(22.5^2 - 3.75^2\right)}{4} = 387 \text{ in}^2$$

$$A_p = \frac{\pi\left(3.75^2\right)}{4} = 11.0 \text{ in}^2$$

$$T_r = 9 \times 0.75 + 2 \times 0.375 = 7.50 \text{ inches}$$

$$S = \frac{22.5^2 - 3.75^2}{4\left(22.5\right)\left(0.75\right)} = 7.29$$

For design checks, AASHTO requires that a 15% deviation from the specified elastomer shear modulus (G) be incorporated, with the least favorable of a higher or lower value for G. The service vertical load on a single bearing is $P_S = P_{DL} + P_{LL} + P_{EQ} = 258 + 40 + 11 = 309$ kips. The 15% reduction in specified shear modulus will be used to compute the shear strain produced by vertical stress.

$$\sigma_s = \frac{309}{387} = 0.798 \text{ ksi}$$

$$\gamma_c = \frac{1.0 \times 0.798}{0.85 \times 0.100 \times 7.29} = 1.288 \text{ rad}$$

$$\gamma_r = \frac{0.375\left(22.5\right)^2\left(0.007\right)}{0.75 \times 7.50} = 0.236 \text{ rad}$$

$$\gamma_{s,s} = {0.75}/{7.5} = 0.100 \text{ rad}$$

$$\gamma_{s,eq} = {8.30}/{7.5} = 1.107 \text{ rad}$$

$$\left(\gamma_c\right)_{static} = 1.288 \text{ rad} \le 3.0 \text{ rad} \rightarrow \text{OK}$$

$$1.75\left(1.288 + 0.236 + 0.100\right) = 2.842 \text{ rad} \le 5.0 \text{ rad} \rightarrow \text{OK}$$

$$\left(1.288 + 0.50 \times 0.236 + 0.100 + 1.107\right) = 2.613 \text{ rad} \le 5.5 \text{ rad} \rightarrow \text{OK}$$

The modeling properties of the elements (usually link-type elements) to be used either in approximate, hand calculations, or in detailed nonlinear structural analysis are determined as follows:

$$k_d = 1.1\frac{100 \times 387}{7.50} = 5,676\frac{\text{lb}}{\text{inch}} = 5.68\frac{\text{kips}}{\text{inch}}$$

$$k_i = \frac{21,750 \times 11.0 + 100 \times 387}{7.50} = 37,200\frac{\text{lb}}{\text{inch}} = 37.2\frac{\text{kips}}{\text{inch}}$$

$$\alpha = \frac{5.58}{37.2} = 0.150$$

$$F_y = \frac{1}{1.0} \times 1.45 \times 11.0 = 16 \text{ kips}$$

$$Q_d = 16(1 - 0.15) = 13.6 \text{ kips}$$

$$D_y = \frac{13.6}{5.58} \cdot \frac{0.15}{1 - 0.15} = 0.43 \text{ inches}$$

$$K_{EFF} = 5.58 + \frac{13.6}{8.30} = 7.22 \text{ kips/inch}$$

$$\xi_{EFF} = \frac{2(13.6)(8.30 - 0.43)}{\pi(8.30)^2(7.22)} = 0.137 \ (13.7\% \text{ effective damping})$$

Yielding of the lead plug essentially adds the effective damping to the system. Response modification for approximate, hand calculation analysis due to effective damping is accounted for in AASHTO by B_L, given in Equation 4.8-17. Response due to effective damping is taken equal to that occurring for 5% damping divided by B_L.

$$B_L = \left(\frac{0.137}{0.05}\right)^{0.30} = 1.353$$

The example is by no means complete, but serves to illustrate sample calculations required for the design of LRB isolators. For complete design examples, refer to the literature (AASHTO, 2014) (Huff, Partial Isolation as a Design Alternative for Pile Bent Bridges in the New Madrid Seismic Zone, 2016) (Huff & Shoulders, Partial Isolation of a Bridge on Interstate 40 in the New Madrid Seismic Zone, 2017).

Basic parameters that define the behavior of FPS systems include:

- μ: dynamic friction coefficient (0.03–0.12, typical)
- R: radius of concave surface
- W: vertical load
- D_{vert}: vertical displacement due to concave sliding surface

A summary of relationships for FPS isolation systems is provided in Equations 4.8-18 through 4.8-24.

$$Q_d = \mu W \tag{4.8-18}$$

$$k_d = \frac{W}{R} \tag{4.8-19}$$

$$k_{eff} = \frac{W}{R} + \mu \cdot \frac{W}{D_{ISO}} \tag{4.8-20}$$

$$\xi_{eff} = \frac{2}{\pi} \cdot \frac{\mu}{\mu + D_{ISO}/R} \tag{4.8-21}$$

$$\varphi = \sin^{-1}\frac{D_{ISO}}{R} \tag{4.8-22}$$

$$D_{vert} = R(1 - \cos\varphi) \cong \frac{D_{ISO}^2}{2R} \tag{4.8-23}$$

$$\mu \leq \frac{D_{ISO}}{R} \leq 0.15 \tag{4.8-24}$$

While not a specification requirement, limits on D_{ISO}/R given in Equation 4.8-24 are generally considered good practice. The lower limit establishes superior re-centering capability. The upper limit assures that small rotation angle assumption about the center of curvature of the concave surface is valid.

Example 4.8-2: Friction Pendulum System (FPS) Bearing for I40 over the Mississippi River

Figure 4.8.5 shows the Hernando de Soto Bridge carrying Interstate 40 over the Mississippi River near Memphis, TN. Figure 4.8.6 is an excerpt from the retrofit plans, which specified the bearing installations at each of the three piers supporting

FIGURE 4.8.5   Hernando de Soto Bridge.

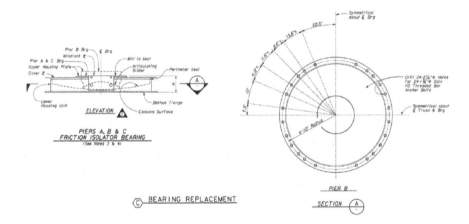

FIGURE 4.8.6   FPS bearing details for the Hernando de Soto Bridge.

the two-span tied arch. Specified physical properties of the 8 ft 10-inch diameter
FPS bearings are as follows:

$\mu = 0.06$
$R = 244$ inches
$W = 12,611$ kips per bearing (center pier); 5,405 kips (arch end piers)
$D_{ISO} = 18.75$ inches (center pier); 27.25 inches (arch end piers)

The derived properties which are useful for structural analysis, either response spectrum or response history, are easily calculated from the given properties.

$$Q_d = 0.06 \times 12{,}611 = 757 \text{ kips, center pier}$$

$$Q_d = 0.06 \times 5{,}405 = 324 \text{ kips, end piers}$$

$$k_d = \frac{12{,}611}{244} = 51.7 \, \frac{\text{kips}}{\text{inch}}, \text{ center pier}$$

$$k_d = \frac{5{,}405}{244} = 22.1 \frac{\text{kips}}{\text{inch}}, \text{ end piers}$$

$$k_{eff} = 51.7 + 0.06 \cdot \frac{12{,}611}{18.75} = 92.1 \frac{\text{kips}}{\text{inch}}, \text{ center pier}$$

$$k_{eff} = 22.1 + 0.06 \cdot \frac{5{,}405}{27.25} = 34.0 \frac{\text{kips}}{\text{inch}}, \text{ end piers}$$

$$\xi_{eff} = \frac{2}{\pi} \cdot \frac{0.06}{0.06 + 18.75 / 244} = 0.279, \text{ center pier}$$

$$\xi_{eff} = \frac{2}{\pi} \cdot \frac{0.06}{0.06 + 27.25 / 244} = 0.223, \text{ end piers}$$

$$\varphi = \sin^{-1} \frac{18.75}{244} = 0.0769 \text{ rad, center pier}$$

$$\varphi = \sin^{-1} \frac{27.25}{244} = 0.1119 \text{ rad, end piers}$$

$$D_{vert} = 244 \left(1 - \cos(0.0769)\right) = 0.722 \text{ inches, center pier}$$

$$D_{vert} = 244 \left(1 - \cos(0.1119)\right) = 1.526 \text{ inches, end piers}$$

$$0.06 \leq \frac{D_{ISO}}{R} = \frac{18.75}{244} = 0.0768 \leq 0.15, \text{ center pier}$$

$$0.06 \le \frac{D_{ISO}}{R} = \frac{27.25}{244} = 0.1117 \le 0.15, \text{ center pier}$$

The maximum seismic force transmitted to each pier is also a simple calculation.

$$F_{max} = Q_d + k_d D_{ISO} = 757 + 51.7 \times 18.75 = 1,726 \text{ kips, center pier}$$

$$F_{max} = Q_d + k_d D_{ISO} = 324 + 22.1 \times 27.25 = 926 \text{ kips, end piers}$$

These calculations were based on weight, W, taken as equal to dead load plus ½ of the live load plus seismic effect. Actual design parameters may be less conservative than this.

Recognize that the design of isolation systems, like most structural design, is iterative in nature. The maximum displacement, D_{ISO}, depends on the physical properties and the derived parameters. The derived parameters depend on the maximum displacement, D_{ISO}.

4.9   COMPUTER MODELING OF BRIDGES

Software packages commonly used for bridge design permit the engineer to perform the required analyses at the various stages of bridge construction: non-composite

FIGURE 4.9.1   Demonbreun Street Viaduct during construction

FIGURE 4.9.2 Sligo Bridge during construction.

properties for deck casting and long- or short-term composite properties for loads applied after the deck has attained the required strength.

Superstructure design is often performed separately from substructure design, although numerous packages make it possible to perform both in the same model.

LRFD-Simon is a free software package for steel girder bridges available from the National Steel Bridge Alliance (NSBA) of the American Institute of Steel Construction (AISC). The program is available from aisc.org/nsba/design-resources. The NSBA also provides a splice design spreadsheet for steel girders, NSBA-Splice, available at the same location.

Free prestressed girder design software is available from the Washington State Department of Transportation as part of their BridgeLink suite of programs. The software may be found at wsdot.wa.gov/eesc/bridge/software.

Regardless of the software used, the engineer needs a clear understanding of the various phases of construction encountered in reaching the finished product, whether a bridge or a building or any other structure. Consider a two-span bridge constructed using precast prestressed concrete girders. The typical method of construction results in girders that act as two simple spans for girder self-weight, wet concrete deck, construction load, and any other load applied prior to the deck reaching design strength. Once the deck reaches design strength, the girders become composite with the deck, and if diaphragm design at the intermediate support is sufficient, act as two continuous spans for loads applied thereafter. The same two-span bridge constructed using steel girders would also behave non-composite prior to deck hardening, but would typically be continuous for all phases of construction. Steel bridge girders are typically built with a continuous section over the intermediate support with adjacent sections installed with field splices.

Figure 4.9.1 was taken during construction of the Demonbreun Street Viaduct in downtown Nashville, TN. The concrete girders are in place spanning from pier to pier. Figure 4.9.2 was taken during construction of State Route 26 over Center Hill Lake and Sligo Road near Smithville, TN. The girder sections over the piers are in place. These pictures illustrate the differences in construction methods for the two girder types. To model either structure with software without consideration of the stages through which the girders are built would be a significant error.

5 Design Ground Motions for Earthquake Engineering of Structures

The focus on seismic design of structures has intensified in the past decade or so. It is important that the structural engineer remain informed as to the nature of the design ground motions for each building code and specification, because differences are significant. The following discussion will focus on three specifications: (1) ASCE 7-16 for buildings, (2) ASCE 43-05 for safety-related nuclear structures, and (3) AASHTO for bridges. Each of these specifications prescribes different levels of ground motion and different spectral measures defining the design ground motions.

First, it will be advisable to become familiar with site characterization, statistical evaluation, directionality, spectral nature, ground motion parameters, ground motion models, and other considerations to appreciate the differences among various modern codes and specifications.

5.1 SITE CHARACTERIZATION

To estimate effects from ground shaking at a particular site, it is necessary to assess the subsurface conditions at that site. Some of the most frequently employed site characteristics are:

- profile depth
- average shear wave velocity (V_{S30}) in the upper 30 m (100 ft)
- depth to achieve a shear wave velocity equal to 1 km/sec ($Z_{1.0}$)
- depth to top of rupture (Z_{TOR})
- depth to achieve a shear wave velocity equal to 2.5 km/sec ($Z_{2.5}$)
- site class (A, B, C, D, E, F).

Profile depth, V_{S30}, and site class are discussed further.

Most design specifications characterize a site based on properties in the upper 30 m (presumably, the reason for this is that the Standard Penetration Test provides information in the upper 30 m only) of the subsurface profile. The Mississippi Embayment of the New Madrid seismic zone possesses sites with much deeper profiles. This raises concern as to whether code-based site factors, which translate bedrock accelerations to the surface, appropriately characterize the seismic hazard at deep soil sites. Researchers have addressed this issue and one of the most comprehensive studies was completed at the University of Illinois (Hashash, Tsai, Phillips,

& Park, 2008). In general, deep sites exhibit a broader, but lower, spectral acceleration plateau, as depicted in Figure 5.1.1.

The shear wave velocity averaged over the upper 30 m of the soil profile is termed
V_{S30}. It is important to recognize that a velocity-based average is not computed the
same way a distance-based average is calculated. The correct calculation of V_{S30} is
given in Equations 5.1-1 and 5.1-2. Also important to recognize is the fact that only
the top 30 m of the profile is used to calculate V_{S30}. Two calculations are necessary:
(1) consider all layers in the top 30 m and (2) consider only the cohesionless layers
in the top 30 m.

$$(V_{S30})_1 = \frac{\sum d_i}{\sum \dfrac{d_i}{V_{Si}}}, \text{all layers} \tag{5.1-1}$$

$$(V_{S30})_2 = \frac{\sum d_i}{\sum \dfrac{d_i}{V_{Si}}}, \text{ include cohesionless layers only} \tag{5.1-2}$$

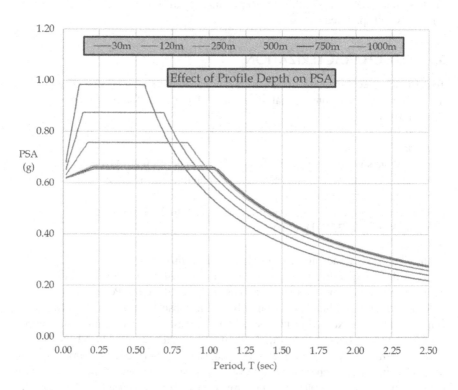

FIGURE 5.1.1   Effect of profile depth on pseudo-spectral acceleration.

Example 5.1-1: Average Shear Wave Velocity Calculation

Suppose a soil profile has been defined as shown in Table 5.1.1. Determine V_{S30} for the site.

First, consider all layers in the top 30 m (100 ft):

$$(V_{S30})_1 = \frac{100}{\frac{12}{810} + \frac{12}{890} + \frac{48}{750} + \frac{28}{985}} = 828 \text{ ft/sec}$$

Next, consider only the cohesionless layers:

$$(V_{S30})_2 = \frac{76}{\frac{48}{750} + \frac{28}{985}} = 822 \text{ ft/sec}$$

Therefore, V_{S30} for this site is 822 ft/sec (251 m/sec).

While measured shear wave velocity testing is clearly preferred, and at times essential, it is possible to get an idea of the V_{S30} value appropriate for a given site. Correlations between topographic slope and V_{S30} have been a subject of research and the OpenSHA (Field, Jordan, & Cornell, OpenSHA: A Developing Community – Modeling Environment for Seismic Hazard Analysis, 2003) Java application permits the determination of "inferred" shear wave velocities, V_{S30}.

Example 5.1-2: Inferred Shear Wave Velocity Determination

Consider, for example, a site in Dyersburg, TN – Interstate 155 over the Mississippi River (Caruthersville Bridge). The coordinates of the site are 36.119°, −86.115°. This is not a plate boundary region, but a "stable continental" region (SCR). The "region type" in OpenSHA should be selected as "Stable Continent" and the methodology set to "Global V_{S30} from Topographic Slope." With the input shown in Figure 5.1.2, the inferred V_{S30}, 299 m/sec, is shown in Figure 5.1.3.

Inferred shear wave velocity is useful as a preliminary guide. It should not be considered a "measured" value, and generally should not be used to establish a definitive shear wave velocity.

ASCE 7-16 (American Society of Civil Engineers, 2017) defines seismic loading for buildings, while AASHTO (AASHTO, 2017) defines seismic loads for bridges.

TABLE 5.1.1
Subsurface Profile for V_{S30} Calculation Example

Layer	d_i, ft	V_s, ft/sec	Type
1	12	810	Cohesive
2	12	890	Cohesive
3	48	750	Cohesionless
4	52	985	Cohesionless
5	78	2750	Cohesionless

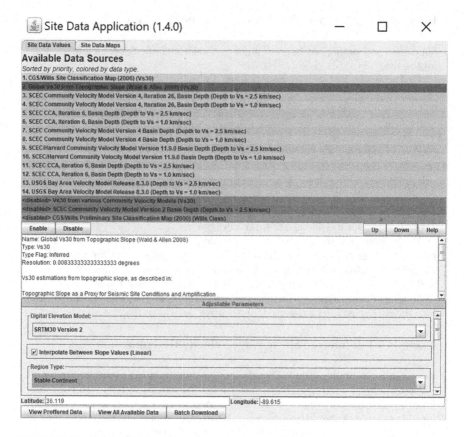

FIGURE 5.1.2   OpenSHA input screen.

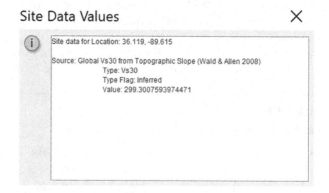

FIGURE 5.1.3   OpenSHA inferred shear wave velocity.

TABLE 5.1.2
Site Class Definitions

Site Class	V_{S30}, ft/sec	N, blows/ft	S_u, psf
A. Hard rock	>5,000	NA	NA
B. Rock	2,500–5,000	NA	NA
C. Very dense soil and soft rock	1,200–2,500	>50	>2,000
D. Stiff soil	600–1,200	15–50	1,000–2,000
E. Soft clay soil	<600	<15	<1,000
F. Soils requiring site response	Liquefaction, peats, highly sensitive or plastic clays		

Site classification is similar, but not identical for these two specifications. In both cases, a particular project is assigned a Site Class. Determination of the Site Class is determined in accordance with Table 5.1.2. The preferred method for Site Class assignment is V_{S30}-based. However, V_{S30} may not be available for many projects. Site Class determination based on the standard penetration test is also possible. Recall that the SPT gives the blow count (N) required to advance a rod into the ground. The calculation of the average N in the upper 30 m follows the same averaging technique as is used for V_{S30}. ASCE 7-16, Section 20.4.2, requires that N be taken no larger than 100 blows/ft for any layer, including any rock layers encountered in the top 30 m. S_u is the undrained shear strength.

Once a site class has been determined, code-based site factors can be determined, and the bedrock accelerations can be translated to the surface.

5.2   GROUND MOTION DIRECTIONALITY

The orientation of ground motion recording instrumentation affects the accelerograms obtained by the instrumentation during strong shaking. Three components are typically recorded – two orthogonal horizontal components and one vertical component. Design ground motions have historically been represented by the geometric mean of the two horizontal components. There are other methods used to condense the intensity of multicomponent horizontal ground shaking into a single parameter.

Most of the original ground motion models were designed to predict spectral acceleration values corresponding to the geometric mean (*GeoMean*) of two horizontal components. Traditional design response spectra, though not explicitly defined as such, have also been typically based on the geometric mean. For each period of vibration, spectra are computed for the two, as-recorded horizontal ground motion components, PSA_{H1} and PSA_{H2}. The design spectral ordinate is then given by Equation 5.2-1 according to the definition of the geometric mean.

$$PSA_{GM} = \sqrt{PSA_{H1} \cdot PSA_{H2}} \qquad (5.2\text{-}1)$$

If instrumentation had been oriented differently, then two different ground motion records would have been obtained during strong ground shaking. For any rotation angle, α, the ground motions obtained can be expressed as given by Equations 5.2-2 and 5.2-3.

$$a_1(\alpha,\, t) = a_X(t)\cos\alpha + a_Y(t)\sin\alpha \qquad (5.2\text{-}2)$$

$$a_2(\alpha,\, t) = -a_X(t)\sin\alpha + a_Y(t)\cos\alpha \qquad (5.2\text{-}3)$$

The rotated components may be computed for each nonredundant rotation angle and the geometric mean spectrum computed. The envelope of all thusly obtained spectra is *GMRotD100*. The median of all thusly obtained spectra is *GMRotD50*.

More recently, design specifications (namely, ASCE 7-16) have required that design spectra be based on the maximum horizontal direction (*RotD100*). This ground motion definition is not a geometric mean, but also relies upon rotation of ground motion components. A single horizontal component is computed according to Equation 5.2-4 for each rotation angle. Spectra are computed for that component across all periods and for all nonredundant rotation angles (up to 180°). The envelope of all those rotated spectra is *RotD100*. The median of all those spectra is *RotD50*. The minimum of those spectra is *RotD00*.

$$a(\alpha,\, t) = a_X(t)\cos\alpha + a_Y(t)\sin\alpha \qquad (5.2\text{-}4)$$

ASCE 7-16 further specifies that the ratio of maximum direction spectra to geometric mean spectra be determined as follows, when conversion between the two bases is required:

- For periods less than or equal to 0.20 sec, *RotD100/GeoMean* = 1.1.
- For a period equal to 1.0 sec, *RotD100/GeoMean* = 1.3.
- For periods greater than or equal to 5.0 sec, *RotD100/GeoMean* = 1.5.
- For periods between those specified above, a linear interpolation is to be used.

The DOS-based Time Series Processing Programs, TSPP (Boore D. M., 2020), are useful in computation of response spectra for rotated ground motion record pairs. For a given recorded record pair, a new record pair could be computed. This new pair is indicative of those which would have been obtained had the instrumentation been oriented differently. The effect of rotation upon the geometric mean response spectra for two record pairs, PEER RSN 0900 from the Landers earthquake and PEER RSN 1147 from the Kocaeli earthquake, is illustrated in Figure 5.2.1. As evident in the figure, the effect may be mild (Kocaeli RSN 1147) or strong (Landers RSN 0900).

Also available in TSPP is the capability to compute various ground motion parameters as well as pseudo-acceleration-spectra, such as the geometric mean, *RotD50*, *RotD100*, and others. The *nga2psa_rot_gmrot* utility provides direct calculation of spectra and ground motion parameters from NGA-formatted ground motion record pairs. Manipulation of spectral data performed in Excel is particularly useful in observing *GeoMean/RotD50* and *RotD100/RotD50* spectral ratios. Figures 5.2.2 and 5.2.3 depict these ratios and their variability over a wide period range for a large number of records. This supports findings in the literature (Boore & Kishida, 2017) that (a) *RotD50* and *GeoMean* are similar, (b) *RotD100* to *GeoMean* ratios are about 1.3–1.4 for periods of 1 sec and more, and (c) the *RotD100* to *GeoMean* ratio is about 1.1–1.2 for short periods. Figures 5.2.4 through 5.2.6 are graphical presentations of

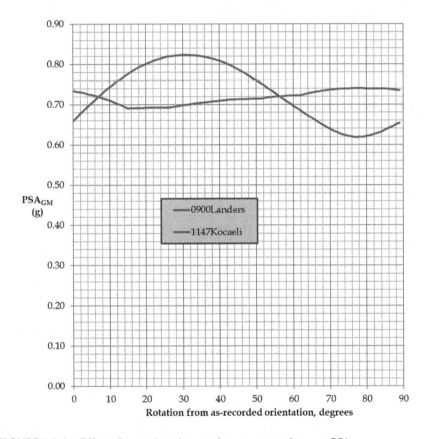

FIGURE 5.2.1 Effect of ground motion rotation on geometric mean *PSA*.

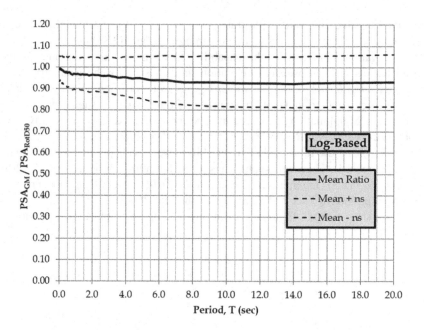

FIGURE 5.2.2   *GeoMean* to *RotD50 PSA* ratio for a ground motion subset.

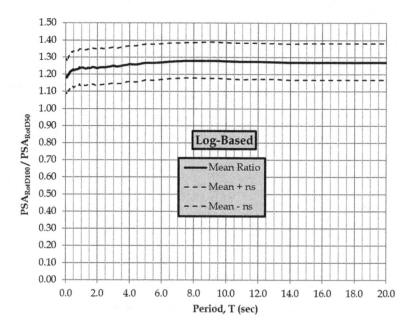

FIGURE 5.2.3   *RotD100* to *RotD50 PSA* ratio for a ground motion subset.

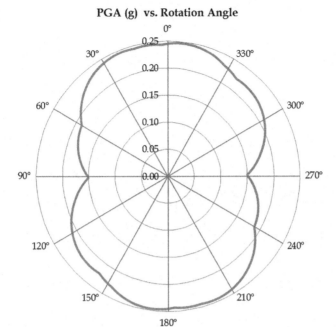

FIGURE 5.2.4 *PGA* versus rotation for Landers RSN 0900.

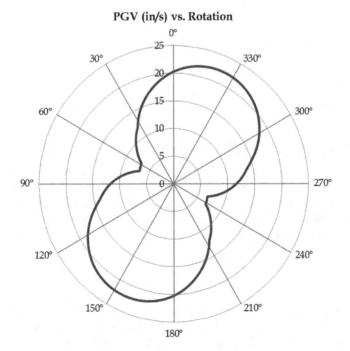

FIGURE 5.2.5 *PGV* versus rotation for Landers RSN 0900.

PGD (inches) vs. Rotation

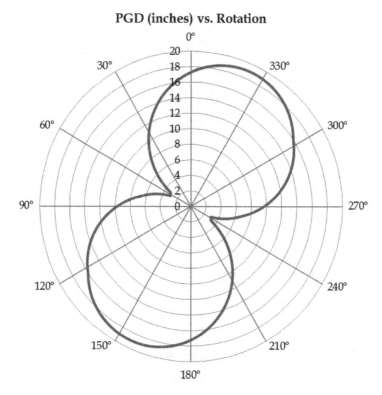

FIGURE 5.2.6   *PGD* versus rotation for Landers RSN 0900.

the effect of instrument rotation on peak ground acceleration (*PGA*), peak ground velocity (*PGV*), and peak ground displacement (*PGD*) for PEER RSN 0900 from the Landers earthquake.

5.3   STATISTICAL CONSIDERATIONS

Design ground motions are typically probabilistic in nature. A Poisson probability distribution, in which past occurrences have no effect on the likelihood of future occurrences, is often assumed. For the Poisson distribution, the probability of exceedance (*PE*), exposure time (*t*), and mean recurrence interval (*MRI*) are related, as given by Equation 5.3-1.

$$PE = 1 - e^{-\frac{t}{MRI}} \qquad (5.3\text{-}1)$$

ASCE 7-16 specifies that design ground motions for buildings are two-thirds of the event producing ground shaking with a 2% *PE* in 50 years. AASHTO specifies that bridges be designed for ground motion having 7% *PE* in 75 years.

Example 5.3-1: Mean Recurrence Interval Calculation

What are the mean recurrence intervals for the design-basis ground motions for buildings and bridges?
  Rearrange the equation to solve for *MRI*:

$$MRI = \frac{-t}{\ln(1-PE)}$$

So, for buildings,

$$MRI = \frac{-50}{\ln(1-0.02)} = 2,475 \text{ years} \approx 2,500 \text{ years}$$

And, for bridges,

$$MRI = \frac{-75}{\ln(1-0.07)} = 1,033 \text{ years} \approx 1,000 \text{ years}$$

It may appear that buildings are designed for a more severe seismic loading than bridges. However, since a factor of two-thirds is applied to the 2%PE/50-Year *PSA* values in ASCE 7-16, this is not necessarily so. Two-thirds of the 2,500-Year *MRI* ground shaking may be greater than, about equal to, or less than the 1,000-Year *MRI* ground shaking. Note that the annual probability of exceedance is the inverse of *MRI*.

A log-normal distribution is typically assumed for many ground motion parameters, including pseudo-spectral acceleration. That is to say, *PSA* is not assumed to be normally distributed in modern design codes, but the natural log of *PSA* is assumed to be normally distributed. Two observations ensue: (1) negative values for median +/− *n*-sigma design values cannot occur, as is possible with the normal distribution, and (2) it becomes possible to compare variability in, say *PSA*, to variability in, say *PSV*, since units become irrelevant.

Example 5.3-2: Log-Normal-Based Demand Calculation

Consider the nonlinear analysis results for structural displacement given in Table 5.3.1. Note that the standard deviation of displacement changes with a change in units, but the standard deviation of the natural log of displacement is independent of units. Using a log-normal distribution, the median + 1σ displacement value becomes:

$$\Delta_{design} = \exp(1.430 + 1.0 \times 0.402) = 6.246 \text{ inches}$$

TABLE 5.3.1

Log-Normal Distribution Calculation Example

i	Displacement, δ (in)	ln(δ)	Displacement, δ (mm)	ln(δ)
1	6.44	1.8625	163.58	5.0973
2	6.04	1.7984	153.42	5.0332
3	3.96	1.3762	100.58	4.6110
4	6.64	1.8931	168.66	5.1279
5	5.51	1.7066	139.95	4.9413
6	5.11	1.6312	129.79	4.8659
7	6.06	1.8017	153.92	5.0365
8	5.54	1.7120	140.72	4.9467
9	3.84	1.3455	97.54	4.5802
10	6.50	1.8718	165.10	5.1066
11	5.90	1.7750	149.86	5.0097
12	2.40	0.8755	60.96	4.1102
13	3.22	1.1694	81.79	4.4041
14	2.24	0.8065	56.90	4.0412
15	3.25	1.1787	82.55	4.4134
16	5.27	1.6620	133.86	4.8968
17	5.98	1.7884	151.89	5.0232
18	3.20	1.1632	81.28	4.3979
19	6.88	1.9286	174.75	5.1634
20	3.50	1.2528	88.90	4.4875
21	2.23	0.8020	56.64	4.0368
22	6.49	1.8703	164.85	5.1050
23	2.22	0.7975	56.39	4.0323
24	6.39	1.8547	162.31	5.0895
25	2.53	0.9282	64.26	4.1630
26	2.51	0.9203	63.75	4.1550
27	2.83	1.0403	71.88	4.2750
28	3.18	1.1569	80.77	4.3916
29	5.98	1.7884	151.89	5.0232
30	3.09	1.1282	78.49	4.3629
σ	1.667	0.402	42.345	0.402
μ	4.498	1.430	114.241	4.664

5.4 GROUND MOTION RESPONSE SPECTRA

Simply put, a "response spectrum" is a spectrum of responses. Applied to earthquake ground motion, response spectra from linear analysis of a single-degree-of-freedom (SDOF) oscillator include:

- spectral displacement, SD
- spectral velocity, SV

- spectral acceleration, *SA*
- pseudo-spectral velocity, *PSV*
- pseudo-spectral acceleration, *PSA*

Response spectra may also be derived based on nonlinear response of the SDOF system. Constant ductility, inelastic displacement spectra are the subject of much research (Chopra, 2005) (Priestley, Calvi, & Kowalsky, 2007) (Huff, 2018) (Huff & Pezeshk, 2016).

For linear analysis-based response spectra, the only required inputs are the ground motion accelerograms and the system damping, expressed as a fraction of critical damping. For constant ductility inelastic spectra, a specified ductility and a hysteretic model are also required. Ductility, in this context, refers to displacement ductility. There are other ductility definitions: rotation, curvature, or strain, for example.

Consider, first, a SDOF system subjected to a ground motion that varies with time, as shown in Figure 5.4.1. The equation of motion for a linear system is given by Equation 5.4-1 and must be solved numerically since the loading history defined by an earthquake ground acceleration is relatively random, rather than being defined by a closed-form function. Structural dynamics theory can be used to demonstrate that the linear structural response, *x*, depends only upon two parameters: the natural period of the system (T) and the fraction (ξ) of critical damping (ξ_{cr}). Damping is energy dissipation that makes a vibrating system eventually stop vibrating.

$$\ddot{x}(t) + 2\xi\omega_n\dot{x}(t) + \omega_n^2 x(t) = -\ddot{x}_g(t) \qquad (5.4\text{-}1)$$

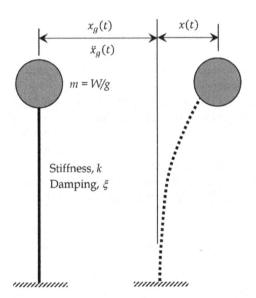

FIGURE 5.4.1 Single-degree-of-freedom oscillator.

$$\omega_n = \sqrt{\frac{k}{m}}, \text{ natural frequency in rad/sec} \tag{5.4-2}$$

$$f_n = \frac{\omega_n}{2\pi}, \text{ natural frequency in cycles/sec (Hz)} \tag{5.4-3}$$

$$T = \frac{1}{f_n}, \text{ natural period in sec} \tag{5.4-4}$$

$$\xi_{cr} = 2m\omega_n = 2\sqrt{km} \tag{5.4-5}$$

Figure 5.4.2 depicts the acceleration history for the East-West component recorded at the "LLO" station during the 2010 $M_w8.8$ Maule, Chile, earthquake. This represents the right side of Equation 5.4-1. SeismoSignal or PRISM may be used to obtain the response history of a linear SDOF system having a natural period of 1.0 sec and 5% damping.

Using PRISM for the solution produces the following maxima:

$$\ddot{x}_{max} = 0.675 \ g$$

$$\dot{x}_{max} = 108 \text{ cm/sec}$$

$$x_{max} = 16.7 \text{ cm}$$

These maxima represent one point on the acceleration, velocity, and displacement spectra plots, respectively.

Spectral acceleration (SA) is the absolute acceleration of an oscillator obtained from a direct solution of the equations of motion.

Spectral velocity (SV) is the relative (to the ground) velocity of an oscillator obtained from a direct solution of the equations of motion.

Spectral displacement (SD) is the relative (to the ground) displacement of an oscillator obtained from a direct solution of the equations of motion. The relative displacement is the displacement of structural interest in building and bridge (or any other structure) design.

Repeating the numerical solution process for a large number of periods and retaining the maximum responses for acceleration, velocity, and displacement produces a plot of the response spectra shown in Figure 5.4.3.

Pseudo-spectral velocity (PSV) is calculated from the spectral displacement (SD) assuming a harmonic relationship:

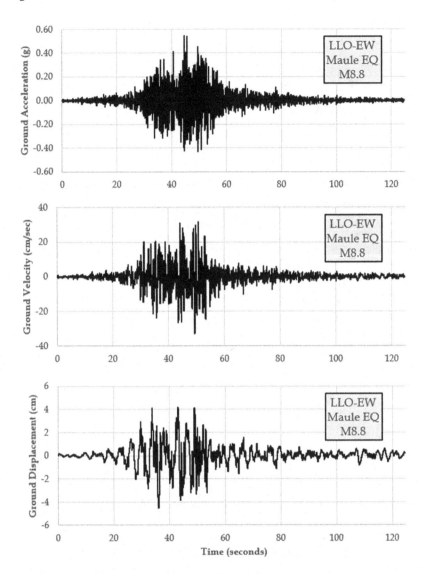

FIGURE 5.4.2 2010 Maule earthquake East-West ground motion at station LLO.

$$PSV = SD \cdot \frac{2\pi}{T} \tag{5.4-6}$$

PSV is often very close to *SV* over a wide range of periods. This is not always the case, however. Figure 5.4.4 is a plot of both *SV* and *PSV* for the Maule ground motion discussed in the previous section.

Pseudo-spectral acceleration (*PSA*) is calculated from the spectral displacement (*SD*) assuming a harmonic relationship:

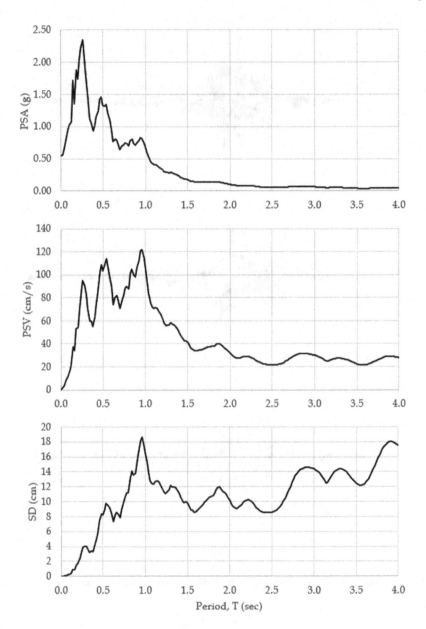

FIGURE 5.4.3   Response spectra for station LLO.

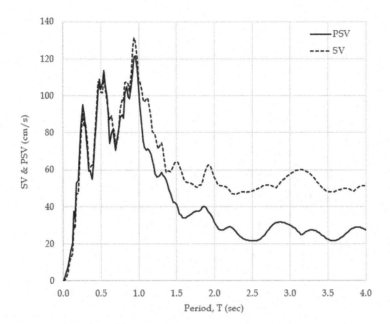

FIGURE 5.4.4 *SV* and *PSV* for station LLO.

$$PSA = SD \cdot \left(\frac{2\pi}{T}\right)^2 \tag{5.4-7}$$

PSA is often very close to *SA* over a wide range of periods. This is not always the case, however. Figure 5.4.5 is a plot of both *SA* and *PSA* for the Maule ground motion discussed in the previous section. The two curves are virtually indistinguishable.

Displacement ductility, μ_D, is defined as maximum displacement divided by yield displacement for a bilinear system. Figure 5.4.6 graphically depicts the maximum displacement, Δ_{INEL}, and the yield displacement, Δ_y. The ratio of maximum displacement, Δ_{INEL}, to that displacement which would have been experienced if nothing yielded, Δ_{EL}, is called the displacement amplification factor, $C\mu$. Periods for which elastic and inelastic displacement are about equal are said to conform to the so-called equal displacement rule.

Figure 5.4.7 shows the inelastic displacement spectra at a displacement ductility value equal to 12 for the Maule ground motion discussed in the previous sections on spectra. The figure reveals that inelastic displacement, SD_{INEL}, may be less than, equal to, or greater than the elastic counterpart, *SD*.

For the seismic design and evaluation of equipment located above ground level in buildings, the ground motion is very different from that applied at the ground. One way to design such components is to include them in the model of the building. Another method could rely upon generating acceleration histories at the level in question from acceleration histories applied at the base of the structure and developing design response spectra (in-structure response spectra, or floor spectra)

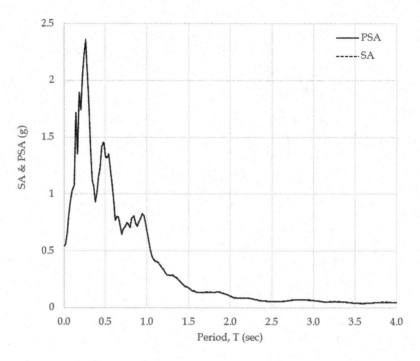

FIGURE 5.4.5   *SA* and *PSA* for station LLO.

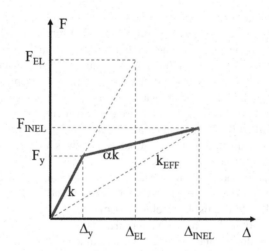

FIGURE 5.4.6   Bilinear force-displacement parameters.

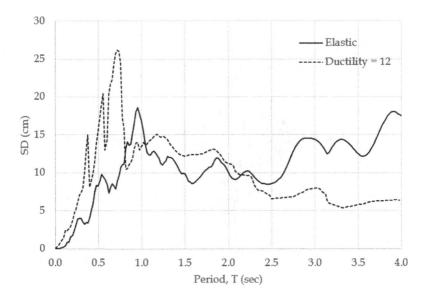

FIGURE 5.4.7   Constant ductility displacement spectra for station LLO.

from those generated acceleration histories. There are methods in the literature that enable direct computation of floor spectra from ground motion spectra (Lucchini, Franchina, & Mollaiolia, 2017).

Example 5.4-1: In-Structure Response Spectra

For an example, consider a one-story building modeled as a SDOF system with a natural period of 1.0 sec and damping equal to 5% of critical. Use PRISM or SeismoSignal to perform a linear analysis of the building subjected to the East-West component of the Maule, Chile (M_W8.8) earthquake recorded at the "LLO" station. Save the resulting roof acceleration history as a new accelerogram to be opened in PRISM or SeismoSignal. Generate the response spectrum from the roof acceleration history. This spectrum represents that which could be used for the design of equipment supported on the roof. Figure 5.4.8 is the roof response history thus obtained. Figure 5.4.9 is the equipment response spectrum.

The in-structure-response spectrum (ISRS) has a clear peak at the natural period of the building, 1.0 sec. For multi-degree-of-freedom systems, these peaks will be evident at multiple periods.

Figure 5.4.10 depicts the roof history from an inelastic analysis of the SDOF, and Figure 5.4.11 depicts the resulting ISRS from nonlinear analysis of the building. The peak at 1 sec still exists. Nevertheless, the ISRS has a much lower amplitude when the building is permitted to respond in the inelastic range.

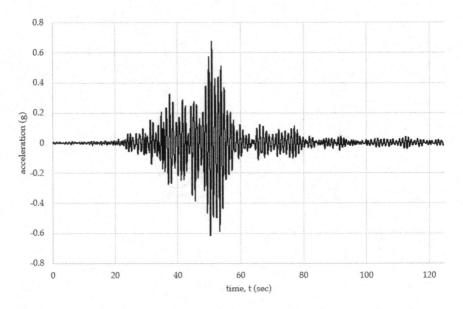

FIGURE 5.4.8 Maule, Chile "LLO" East-West linear response history ($T = 1$ sec).

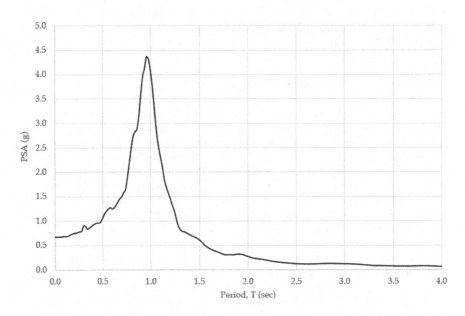

FIGURE 5.4.9 Maule, Chile "LLO" East-West linear elastic ISRS.

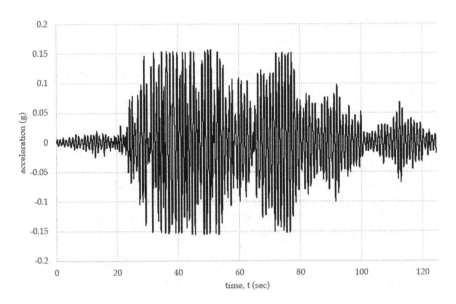

FIGURE 5.4.10   Maule, Chile "LLO" East-West inelastic response history (T = 1 sec).

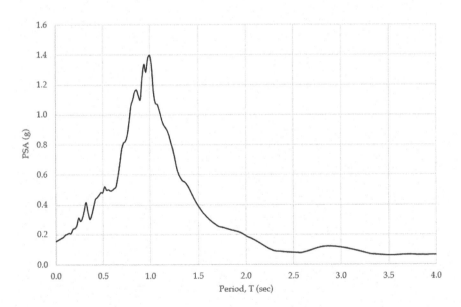

FIGURE 5.4.11   Maule, Chile "LLO" East-West nonlinear ISRS (T = 1 sec).

For the generation of in-structure-response spectra, ASCE 7-16 outlines specific requirements:

1. A minimum of three natural modes of vibration in two orthogonal, horizontal directions must be used.
2. The natural periods and modal participation factors for each considered mode shall be determined.
3. A component dynamic amplification factor, DAF, is computed for each natural period of the building structure, T_x. T_p is the natural period of the component in question. Figure 5.4.12 is based on Section 13.3 of ASCE 7-16 and is used to determine the D_{AF} for each structure period.
4. For each mode in each direction, compute the modal acceleration, A_{ix}, at each floor of the building structure. The parameters in Equation 5.4-8 are as follows: (a) p_{ix} is the modal participation factor for mode i multiplied by the mode shape value at floor x, (b) S_{ai} is the spectral acceleration from the structure design response spectrum at the period corresponding to mode i, and (c) D_{AF} is the amplification factor from Figure 5.4.12.

$$A_{ix} = p_{ix} \cdot S_{ai} \cdot D_{AF} \qquad (5.4\text{-}8)$$

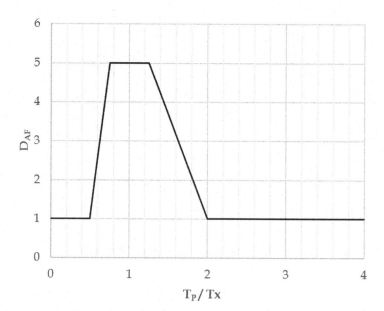

FIGURE 5.4.12   Component dynamic amplification factor.

TABLE 5.4.1
Modal Properties for ISRS Example

Mode	T, sec	PSA, g	Γ	Mode Shapes			
				ϕ_1	ϕ_2	ϕ_3	ϕ_4
1	1.7284	0.347	1.5670	0.07161	0.16383	0.68349	0.70772
2	0.4305	0.750	−1.1579	−0.52569	0.83990	0.0−521	0.12317
3	0.2708	0.750	0.0430	0.05503	0.02613	−0.36583	0.92869
4	0.2075	0.750	−0.3602	−0.76824	0.63947	−0.02402	0.01748

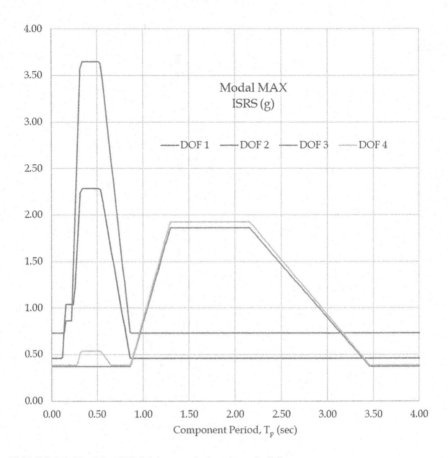

FIGURE 5.4.13 Max ISRS for example four-story building.

5. The ISRS at each period is taken as the "maximum floor acceleration" for at least the first three modes, but no less than the structure design response spectral ordinate at each period.

This still requires a good deal of interpretation.

Example 5.4-2: Multi-Degree-of-Freedom In-Structure-Response Spectra

Consider a three-story building with the dynamic properties indicated in Table 5.4.1. Taking a literal interpretation of the stated requirements – the maximum of all considered modes – results in Figure 5.4.13. Combining modal responses using the square-root-of-sum-of-squares (SRSS) for each mode results in Figure 5.4.14. While the two are similar, there are differences. It is not clear if a modal combination is required.

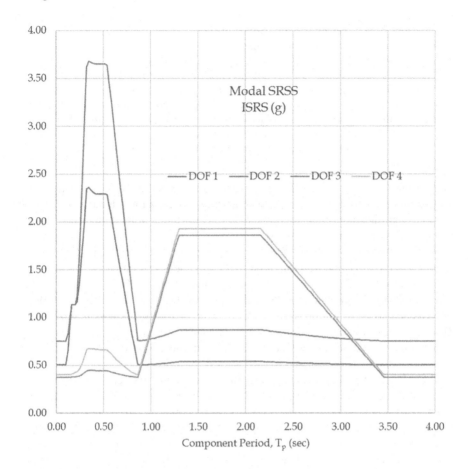

FIGURE 5.4.14 SRSS ISRS for example four-story building.

5.5 GROUND MOTION PARAMETERS

Peak ground acceleration (*PGA*), peak ground velocity (*PGV*), and peak ground displacement (*PGD*) are the three most basic ground motion parameters used as indicators of earthquake damage potential.

PGA is relatively easily estimable from the recorded ground motion. *PGV* may be obtained by integrating (numerically) the ground acceleration history, and *PGD* may be obtained by integrating (numerically) the ground velocity history. Due to "noise" in ground motion records and the need for baseline adjustment and filtering to remove this "noise," *PGV* and *PGD* are relatively difficult to pinpoint. *PGV* and particularly *PGD* estimates are highly dependent upon the baseline adjustment and filtering techniques used to remove signal "noise."

Example 5.5-1: Baseline Adjustment and Filtering of Ground Motion Records

Take, as an example, the North-South component of the ground motion recorded at the "IBR013" station during the 2011 M_W9.0 Japan earthquake. Table 5.5.1 summarizes *PGA, PGV,* and *PGD* values obtained using three different baseline adjustment and filtering schemes available with PRISM. It is clear from the table that estimated *PGA, PGV,* and *PGD* values from ground motion recordings are highly dependent upon the processing scheme applied to the raw record.

Significant duration is one method used to measure the strong motion duration for a ground motion record. While an accelerogram may have a total duration of 300 sec or more, significant duration attempts to define how much of the total accelerogram consists of "strong" ground shaking.

Significant duration may be D_{5-75}, the time between accumulation of 5% of the total Arias Intensity and accumulation of 75% of the Arias Intensity. On the other hand, it may be D_{5-95}, the time between accumulation of 5% of the total Arias Intensity and accumulation of 95% of the Arias Intensity. In fact, significant duration can be computed for any percentages of Arias Intensity accumulation. D_{5-75} and D_{5-95} are the most popular definitions of significant duration.

Significant duration is discussed further in a subsequent section on Arias Intensity, where an example is presented.

Bracketed duration is defined as the time between the first and last occurrences of a particular ground acceleration value for a given accelerograms. For example, bracketed duration with a threshold acceleration equal to 0.10 would be the time between the first occurrence of 0.10 g in the accelerograms and the last occurrence of 0.10 g in the accelerograms. As positive and negative accelerations have little physical significance, the bracketed duration is often computed by squaring the ordinates of the accelerograms and searching for the first and last occurrences of the square of the threshold acceleration.

Cumulative absolute velocity (*CAV*) is one ground motion parameter that has been suggested as being indicative of structural damage potential, and the expression for *CAV* is given in Equation 5.5-1. Notice that *CAV* has units of velocity.

TABLE 5.5.1

Effect of Processing on *PGA, PGV,* and *PGD* – Station IBR013 (Tohoku)

Baseline Correction	Second Order	Second Order	Second Order
Causal Filter Type	Butterworth	Butterworth	Butterworth
Filtering	Fourth-order bandpass	Second-order bandpass	Fourth-order bandpass
Frequencies	0.02–40 Hz	0.02–40 Hz	0.05–40 Hz
PGA, g	1.368	1.383	1.398
PGV, cm/sec	95	102	76
PGD, cm	166	214	49

$$CAV = \int_{0}^{t_{max}} |a(t)| \, dt \qquad (5.5\text{-}1)$$

Arias intensity is a ground motion parameter, suggested by Chilean engineer Arturo Arias in 1970, which has been suggested as indicative of landslide potential. Notice that I_A has units of velocity in Equation 5.5-2.

$$I_A = \frac{\pi}{2g} \int_{0}^{t_{max}} [a(t)]^2 \, dt \qquad (5.5\text{-}2)$$

Figure 5.5.1 shows the Arias Intensity accumulation with time for the North-South component of ground motion recorded at station "CHY116" (PEER RSN 1250) during the $M_W7.62$ Chi-Chi Taiwan earthquake in 1999. The significant duration, $D_{5\text{-}95}$, is indicated in the figure and is equal to 129 sec for this record. Figure 5.5.2 shows the bracketed duration for a threshold acceleration equal to 0.05 g for the same record. Zooming in on Figure 5.5.2 would reveal a value for the bracketed duration of about 0.50 sec. So quite a large difference can exist between significant and

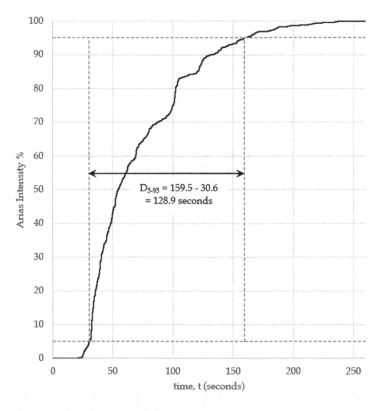

FIGURE 5.5.1 Arias intensity with $D_{5\text{-}95}$ (PEER RSN 1250-NS).

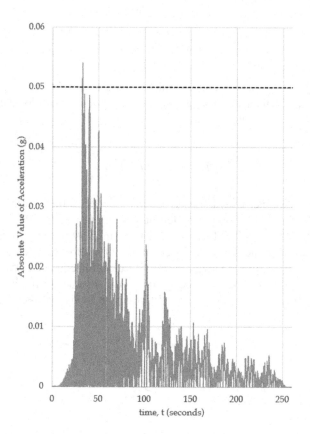

FIGURE 5.5.2   Bracketed duration (0.05g) (PEER RSN 1250-NS).

bracketed durations. A record possessing high values for both the bracketed duration and the significant duration would seem to be indicative of one having a high damage potential.

Specific energy density (*SED*) is one of the few ground motion parameters computed from the ground velocity history, rather than the ground acceleration history. *SED* has units of length^2 over time, given by Equation 5.5-3:

$$SED = \int_0^{t_{max}} \left[v(t) \right]^2 dt \qquad (5.5-3)$$

Each of the parameters – PSV_{Max} / PSA_{Max}, PGD / PGV, and SD_{Max} / PSV_{Max} – has units of time, usually in seconds. These parameters have been proposed as indicative of breakpoints on smoothed design response spectra.

Figure 5.5.3 is an example of a smoothed design response spectrum, with control periods T_o, T_S, and T_L indicated on the figure. Various multiples of the parameters discussed in this section have been proposed as estimates of the transition

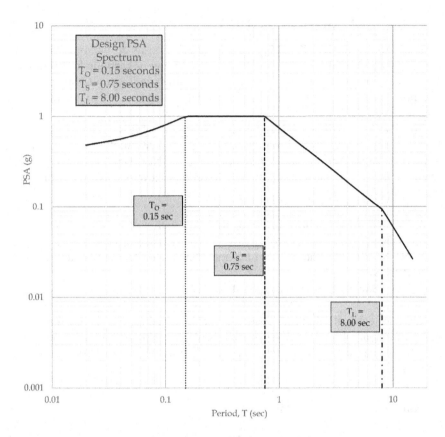

FIGURE 5.5.3   Sample design response spectrum.

periods indicated (Trombetti, Silvestri, Gasparini, Righi, & Ceccoli, 2008) (Bommer, Elnashai, & Weir, 2000). Some of the proposed equations are given here in Equations 5.5-4 through 5.5-7:

$$T_L = 8 \cdot \frac{PGD}{PGV} \qquad\qquad (5.5\text{-}4)$$

$$T_L = 20 \cdot \frac{PGV}{PGA} \qquad\qquad (5.5\text{-}5)$$

$$T_L = 2\pi \cdot \frac{SD_{Max}}{PSV_{Max}} \qquad\qquad (5.5\text{-}6)$$

$$T_S = 2\pi \cdot \frac{PSV_{Max}}{PSA_{Max}} \qquad\qquad (5.5\text{-}7)$$

5.6 GROUND MOTION MODELS

Ground motion models (GMM, formerly referred to as attenuation equations) predict some ground motion parameter as a function of earthquake magnitude, subsurface conditions at the site in question, fault type, source-to-site distance, and possibly many additional parameters, depending on the complexity of the model.

The Pacific Earthquake Engineering Research (PEER) center has been a leader in developing ground motion models used in seismic hazard assessment. The umbrella project for model development is the Next Generation Attenuation (NGA) project. Ground motion models have been developed through the NGA project for:

1. NGA-West2 – Shallow crustal earthquakes within active tectonic regions (ATR)
2. NGA-East – Earthquakes within the Central and Eastern United States (CEUS)
3. NGA-Sub – Subduction zone earthquakes

In addition to Excel "flatfiles" containing large metadata, Excel files that enable ground motion parameter estimation are available at PEER, for NGA-West2 and NGA-East, and at the UCLA B. John Garrick Institute for the Risk Sciences for NGA-Sub.

Example 5.6-1: Ground Motion Models

Figure 5.6.1 was obtained from the NGA-East GMM spreadsheet for three scenarios – (a) a magnitude M_W7.55 earthquake with a source-to-site distance equal to 55 km, (b) a magnitude M_W 6.3 earthquake with a source-to-site distance equal to 30 km, and (c) a magnitude 5.7 earthquake with a source-to-site distance equal to 5 km. The NGA-East GMM predicts ground motion parameters for subsurface conditions characterized by a shear wave velocity, V_{S30} = 3,000 m/s. Figure 5.6.1 illustrates an important concept: small-magnitude events close to a site may control short-period response, while large magnitude distant events control longer period response. This is certainly not a rule for all sites, but neither is it an uncommon occurrence.

A more recent model for shallow crustal earthquakes is available in the literature (Graizer & Kalkan, 2015) and includes estimates of mean (geometric mean) and maximum direction (RotD100) spectra as well as estimates of log-based standard deviation.

The USGS incorporates many GMMs into the national seismic hazard-mapping project upon which many design codes and specifications are based. Logic trees are used to establish probabilistic ground motion parameters for design. Ground motion models included in one of the logic trees from the 2014 version of the project (Petersen, et al., 2014) are shown in Figure 5.6.2 for the Central and Eastern United States (CEUS). Several other logic trees may be found in the report.

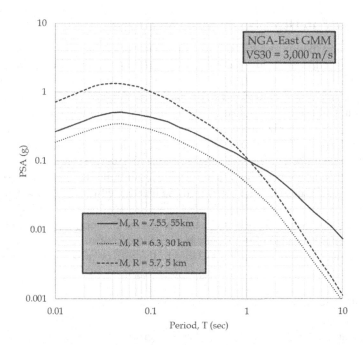

FIGURE 5.6.1   NGA-East GMM scenario spectra.

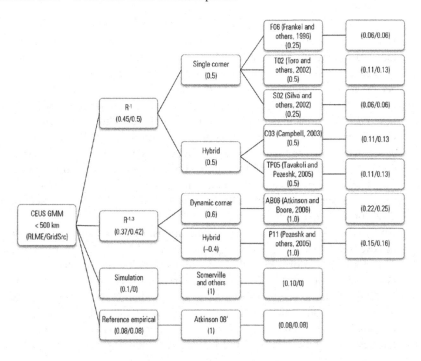

FIGURE 5.6.2   USGS 2014 Logic Tree – CEUS ($R < 500$ km) (Petersen, et al., 2014).

5.7   ASCE 7-16 DESIGN GROUND MOTIONS

Design ground motions in ASCE 7-16 (American Society of Civil Engineers, 2017) are risk-targeted, maximum direction based (*RotD100*). Distinction is made between MCE_G, the maximum considered earthquake geometric mean acceleration, and MCE_R, the maximum considered earthquake risk-targeted *RotD100* acceleration. MCE_G is not adjusted for targeted risk. It is uniform hazard and represents the geometric mean of two horizontal components. MCE_R is adjusted for targeted risk, and is *RotD100* based. MCE_G accelerations are used for liquefaction assessment, lateral spreading, seismic settlement, and other soil-related issues. MCE_R is used in developing design response spectra for structural analysis of buildings and other structures.

ASCE 7-16 defines near-fault sites as those meeting either of the following:

- The site is within 9.5 miles (15 km) of the surface projection of a known active fault capable of producing M_W7 or larger events.
- The site is within 6.25 miles (10 km) of the surface projection of a known active fault capable of producing M_W6 or larger events.

Near-fault sites are more likely to be subjected to "pulse-type" ground motions (a noticeable pulse is visible in the velocity history). The PEER Ground Motion database has capabilities for identifying pulse-type ground motion records.

The procedure for generating design response spectra in accordance with ASCE 7 is outlined below.

1. Determine uniform hazard, *RotD100*-based (maximum direction) subsurface, rock spectral accelerations $S_{S\text{-}UHS}$ (spectral acceleration at a period of 0.20 sec) and $S_{1\text{-}UHS}$ (spectral acceleration at a period of 1.0 sec). The accelerations should correspond to those having 2% probability of exceedance in 50 years (about a 2,500-year mean recurrence interval). *GeoMean*-based accelerations may be obtained from the USGS Unified Hazard Tool. *GeoMean*-based accelerations may be converted to *RotD100*-based accelerations by applying factors of (a) 1.1, for periods equal to or less than 0.20 sec, (b) 1.3, for a period of 1.0 sec, (c) 1.5, for periods equal to or greater than 5.0 sec, and (d) linear interpolation for periods between 0.20 and 1.0 sec and for periods between 1.0 and 5.0 sec. Refer to ASCE 7-16, Section 21.2, for the above factors.

2. Apply risk coefficients, C_R, to the accelerations. Risk coefficients are found in ASCE 7-16, Figure 22-18A (C_{RS}) and ASCE 7-16, Figure 22-19A (C_{R1}). This converts uniform hazard accelerations to risk-targeted accelerations. For periods other than 0.20 and 1.0 sec, risk coefficients are to be determined as follows. At periods equal to or less than 0.20 sec, C_R is equal to C_{RS}. At periods equal to or greater than 1.0 sec, C_R is equal to C_{R1}. At periods between 0.20 and 1.0 sec, a linear interpolation is used. The risk-targeted accelerations at 0.20 and 1.0 sec are S_S and S_1, respectively.

3. Unless site-specific hazard analysis and site response analysis is required, apply site factors from ASCE 7-16, Chapter 11, F_a and F_v. $S_{MS} = F_a \times S_S$ and $S_{M1} = F_v \times S_1$.
4. Design response spectrum key points are defined by $S_{DS} = 2/3 \times S_{MS}$ and $S_{D1} = 2/3 \times S_{M1}$. Control periods are $T_S = S_{D1} / S_{DS}$, $T_o = 0.20T_S$, and T_L as taken from maps in ASCE 7-16, Chapter 22. Figure 5.7.1 depicts a generic design response spectrum with the control points identified.

Site factors F_a and F_v are presented in Tables 5.7.1 and 5.7.2, taken directly from ASCE 7-16, with subsequent table notes regarding site-specific requirements.

A site response analysis in accordance with ASCE 7-16, Section 21.1, is required for Structures on Site Class F sites, unless exempted in accordance with Section 20.3.1. A ground motion hazard analysis shall be performed in accordance with Section 21.2 for the following:

- Seismically isolated structures and structures with damping systems on sites with $S_1 \geq 0.60$
- Structures on Site Class E sites with $S_S \geq 1.00$
- Structures on Site Class D and E sites with $S_1 \geq 0.20$

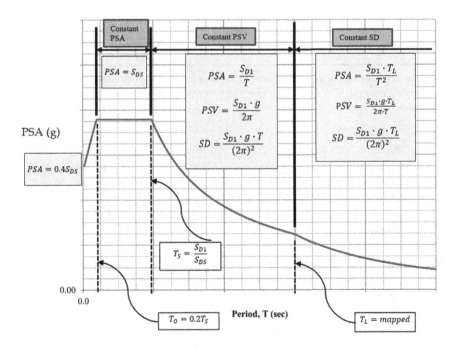

FIGURE 5.7.1 ASCE 7-16 design response spectrum shape.

TABLE 5.7.1
ASCE 7-16 Site Factor F_a

Site Class	$S_s \leq 0.25$	$S_s = 0.50$	$S_s = 0.75$	$S_s = 1.00$	$S_s = 1.25$	$S_s \geq 1.50$
A	0.80	0.80	0.80	0.80	0.80	0.80
B	0.90	0.90	0.90	0.90	0.90	0.90
C	1.30	1.30	1.20	1.20	1.20	1.20
D	1.60	1.40	1.20	1.10	1.00	1.00
E	2.40	1.70	1.30	See 11.4.8	See 11.4.8	See 11.4.8
F	See 11.4.8	See 11.4.8	See 11.4.8	See 11.4.8	See 11.4.8	See 11.4.8

TABLE 5.7.2
ASCE 7-16 Site Factor F_v

Site Class	$S_1 \leq 0.10$	$S_1 = 0.20$	$S_1 = 0.30$	$S_1 = 0.40$	$S_1 = 0.50$	$S_1 \geq 0.60$
A	0.80	0.80	0.80	0.80	0.80	0.80
B	0.80	0.80	0.80	0.80	0.80	0.80
C	1.50	1.50	1.50	1.50	1.50	1.40
D	2.40	2.20[a]	2.00[a]	1.90[a]	1.80[a]	1.70[a]
E	4.20	3.30[a]	2.80[a]	2.40[a]	2.20[a]	2.00[a]
F	See 11.4.8	See 11.4.8	See 11.4.8	See 11.4.8	See 11.4.8	See 11.4.8

Exception: A ground motion hazard analysis is not required for structures other than seismically isolated structures and structures with damping systems where (interpretation of author Huff for the second bulleted item):

- Site Class E sites with $S_s \geq 1.00$, provided F_a is taken as that for Site Class C
- Site Class D sites with $S_1 \geq 0.20$, provided $T_s = S_{D1}/S_{DS}$, $T_o = 0.13S_{D1}/S_{DS}$, F_v = 1.5 times the tabulated value for Site Class D
- Site Class E with $S_1 \geq 0.20$, provided $T \leq T_s$ and the equivalent lateral force procedure is used

When subsurface conditions are not well enough defined to establish the appropriate Site Class, Site Class "D-Default" is used. In such cases, F_a is not to be taken less than 1.2, regardless of the tabulated value.

Ground motion selection and modification criteria for linear response history analysis (LRHA) specified in ASCE 7-16, Section 12.9.2.3, are summarized below.

- No fewer than three pairs of "spectrally matched" orthogonal components are to be used.

- The target response spectrum is the design response spectrum, based on S_{DS} and S_{D1}.
- The period range of interest for spectrum matching is $0.8T_{LOWER}$ to $1.2T_{UPPER}$. T_{LOWER} is defined as the period at which 90% of the total system mass has been accounted for in each horizontal direction in a modal dynamic analysis. T_{UPPER} is defined as the larger of the two orthogonal, horizontal fundamental periods of vibration from a frequency analysis of the structure.
- Within the period range of interest, the suite mean 5% damped *PSA* spectrum is to be within 10% of the design response spectrum.

Ground motion selection and modification criteria for nonlinear response history analysis (NLRHA) specified in ASCE 7-16, Chapter 16, are summarized next.

- The target response spectrum is to be the MCE_R spectrum, defined by S_{MS} and S_{M1} (not the "design" response spectrum defined by S_{DS} and S_{D1}).
- No fewer than 11 ground motion record pairs selected from events within the general tectonic regime and having generally consistent magnitudes and fault distances as those controlling the target spectrum shall be developed for the analyses. Records shall have similar spectral shape to the target.
- Simulated ground motions are permitted when the required number of ground motion records is not available.
- Ground motions may be amplitude scaled or spectrally matched. Spectral matching shall not be used for near-fault sites unless the pulse-type character of the ground motion is retained in the matching process.
- The upper bound on the period range of interest is required to be equal to twice the larger of the two fundamental periods in the two orthogonal, horizontal directions. The argument may be made that this should be extended to 2.5 times the larger period. For bridge structures using a conventional seismic design strategy, displacement ductility values up to six are permitted. This results in an effective period, based on secant stiffness, equal to 2.45 (square root of 6) times the period calculated from initial stiffness for an elastic-perfectly-plastic system.
- The period range of interest shall have a lower bound equal to the smaller of (a) the period required to achieve at least 90% mass participation in each horizontal, orthogonal direction and (b) 20% of the smallest fundamental period in the two horizontal, orthogonal directions.
- For amplitude scaling, record pairs shall be scaled such that the suite mean *RotD100*-based *PSA* spectrum generally matches or exceeds the MCE_R target and does not fall below 90% of the MCE_R target spectrum at any period within the range of interest.
- For spectral matching, record pairs shall be modified such that the average *RotD100* spectrum for the suite equals or exceeds 110% of the MCE_R target across the period range of interest.
- For near-fault sites, records should be rotated to fault-normal (*FN*) and fault-parallel (*FP*) orientations for application in structural analysis.

- Where vertical ground motion is of concern, see ASCE 7-16, Section 11.9. Vertical spectra are required in NLRHA when (a) vertical elements of the gravity force resisting system are discontinuous and (b) for non-building structures designated in ASCE 7-16, Chapter 15.

Example 5.7-1: ASCE 7-16 Design Ground Motion

Consider a site near Memphis, TN, with coordinates 35°09'11" N, 90°03'07"W (+35.153, -90.052). V_{S30} for the site is 200 m/sec, corresponding to Site Class D (Default) subsurface conditions. Use the USGS Unified Hazard Tool to determine uniform hazard, *GeoMean*-based accelerations, S_{S-UHS} and S_{1-UHS}. For this example, use the "2014 v4.0.x" data set for spectral acceleration and "2014 v4.1.4" data set for disaggregation (disaggregation for the "2014 v4.0.x" data set was not available at the time of the work presented in this example). From the USGS Unified Hazard Tool:

$$S_{S-UHS} = 1.0692\,g$$

$$S_{1-UHS} = 0.3073\,g$$

From ASCE 7-16, Figures 22-18A and 22-19A:

$$C_{RS} = 0.875$$

$$C_{R1} = 0.873$$

Recall that USGS spectral accelerations are *GeoMean*-based, while ASCE 7-16 requires spectra based on *RotD100*. So, application of the conversion factors (1.1 for S_S and 1.3 for S_1), in addition to application of the risk factors, is required to obtain the MCE_R accelerations.

$$S_S = 1.0692 \times 1.1 \times 0.875 = 1.0291\,g$$

$$S_1 = 0.3073 \times 1.3 \times 0.873 = 0.3488\,g$$

Site factor, F_a, from ASCE 7-16 is interpolated from tabulated values:

$$F_a = 1.1 - 0.1\left(\frac{1.0291-1}{1.25-1}\right) = 1.0884$$

However, since Site Class "D-Default" was used for this site, take:

$$F_a = 1.2$$

For F_v the tabulated value again requires interpolation:

$$F_v = 2.0 - 0.1\left(\frac{0.3488-0.3}{0.4-0.3}\right) = 1.9512$$

However, since the Site Class is D and $S_1 = 0.3488 > 0.20$, take:

$$F_v = 1.5 \times 1.9512 = 2.9268$$

Therefore, the control points for the design response spectrum are:

$$S_{MS} = 1.2 \times 1.0291 = 1.2349$$

$$S_{M1} = 2.9268 \times 0.3488 = 1.0209 \ g$$

$$S_{DS} = \frac{2}{3} \times 1.2 \times 1.0291 = 0.8233 \ g$$

$$S_{D1} = \frac{2}{3} \times 2.9268 \times 0.3488 = 0.6806 \ g$$

$$T_S = \frac{0.6806}{0.8233} = 0.8267 \text{ sec}$$

$$T_o = \frac{0.20}{1.5} \times 0.8267 \text{ sec} = 0.1102 \text{ sec}$$

Finally, from ASCE 7-16, Figure 22-14:

$$T_L = 12 \text{ sec}$$

For sites like Memphis in the Mississippi Embayment (ME) of the New Madrid seismic zone (NMSZ), code-based site factors may not be the best choice for design. Particularly for the ME in the NMSZ, research on depth-dependent site factors obtained from site response analyses have been published (Malekmohammadi & Pezeshk, 2014; Hashash, Tsai, Phillips, & Park, 2008). The published tables are repeated here in Tables 5.7.3 and 5.7.4 (Hashash, Tsai, Phillips, & Park, 2008) and in Tables 5.7.5, 5.7.6, 5.7.7, and 5.7.8 (Malekmohammadi & Pezeshk, 2014). Presumably, there is a typographical error in the Malekmohammadi tables for F_v, as dependence of this factor is on S_1, not S_S. Replace S_S with S_1 in the header row for the Malekmohammadi tables for F_v.

For projects within the ME, it would be advisable to examine the site-specific, depth-dependent site factor and subsequent effect on the design response spectrum. For the example above, the ASCE 7-16 site factor, F_v, amplified by the 1.5 factor was shown to produce a more severe design response spectrum than that obtained by applying the depth-dependent NMSZ site factors. Hence, the ASCE 7 procedure was retained. This will not always be the case. When required by the governing design authority, a site-specific site response analysis procedure may be required in lieu of either code-based or research-based site factors.

NMSZ-specific site amplification factors make a distinction between the so-called uplands and lowlands regions. The lowlands region generally is characterized by lower shear wave velocities in the upper 70 m or so of the subsurface profile.

TABLE 5.7.3

Site Factor F_a from Hashash, Tsai, Phillips, & Park, 2008

Thickness (m)	$S_S = 0.25$		$S_S = 0.50$		$S_S = 0.75$		$S_S = 1.00$		$S_S \geq 1.25$	
	Up	Low	Up	Low	Up	Low	Up	Low	Up	Low
30	1.46	1.41	1.32	1.27	1.18	1.13	1.11	1.06	1.06	1.01
100	1.41	1.31	1.27	1.17	1.13	1.03	1.06	0.96	1.01	0.91
200	1.36	1.21	1.22	1.07	1.08	0.93	1.01	0.86	0.96	0.81
300	1.31	1.11	1.17	0.97	1.03	0.83	0.96	0.76	0.91	0.71
500	1.27	1.06	1.13	0.92	0.99	0.78	0.92	0.71	0.87	0.66
1000	1.23	1.04	1.09	0.90	0.95	0.76	0.88	0.70	0.83	0.64

Up = uplands ; Low = lowlands.

TABLE 5.7.4

Site Factor F_v from Hashash, Tsai, Phillips, & Park, 2008

Thickness (m)	$S_1 = 0.10$		$S_1 = 0.20$		$S_1 = 0.30$		$S_1 = 0.40$		$S_1 \geq 0.50$	
	Up	Low	Up	Low	Up	Low	Up	Low	Up	Low
30	2.40	2.40	2.00	2.00	1.80	1.80	1.60	1.60	1.50	1.50
100	2.70	2.55	2.30	2.15	2.10	1.95	1.95	1.75	1.80	1.65
200	2.85	2.67	2.45	2.27	2.25	2.07	2.08	1.87	1.91	1.77
300	2.95	2.77	2.55	2.37	2.37	2.17	2.18	1.97	2.01	1.87
500	3.00	2.82	2.60	2.42	2.42	2.22	2.23	2.02	2.06	1.92
1000	3.05	2.87	2.65	2.47	2.47	2.27	2.28	2.07	2.08	1.97

Up = uplands; Low = lowlands.

TABLE 5.7.5

Uplands Site Factor F_a from Malekmohammadi & Pezeshk, 2014

V_{S30}, m/s	Site Class	Depth, m	$S_S \leq$ 0.25	$S_S = 0.50$	$S_S = 0.75$	$S_S = 1.00$	$S_S \geq 1.25$
560	C	30	1.509	1.228	1.049	0.923	0.829
		70	1.624	1.285	1.081	0.940	0.837
		140	1.618	1.250	1.036	0.892	0.788
		400	1.362	1.011	0.819	0.693	0.604
		750	1.069	0.774	0.617	0.517	0.447
270	D	30	1.528	1.057	0.803	0.647	0.543
		70	1.660	1.117	0.836	0.666	0.553
		140	1.667	1.095	0.807	0.637	0.525
		400	1.421	0.896	0.646	0.501	0.408
		750	1.123	0.691	0.491	0.377	0.304
180	E	30	1.451	0.900	0.638	0.490	0.396
		70	1.581	0.954	0.666	0.505	0.405
		140	1.592	0.938	0.645	0.485	0.385
		400	1.362	0.771	0.518	0.383	0.301
		750	1.079	0.595	0.394	0.288	0.225

TABLE 5.7.6

Uplands Site Factor F_v from Malekmohammadi & Pezeshk, 2014

V_{S30}, m/s	Site Class	Depth, m	$S_1 \leq$ 0.10	$S_1 =$ 0.20	$S_1 = 0.30$	$S_1 = 0.40$	$S_1 \geq 0.50$
560	C	30	3.304	2.841	2.550	2.340	2.179
		70	4.428	3.862	3.496	3.227	3.017
		140	5.630	4.947	4.498	4.165	3.904
		400	4.171	3.708	3.394	3.158	2.971
		750	3.559	3.181	2.921	2.724	2.567
270	D	30	3.771	2.753	2.176	1.803	1.543
		70	4.383	3.245	2.586	2.155	1.853
		140	4.974	3.711	2.970	2.483	2.140
		400	4.397	3.318	2.674	2.246	1.943
		750	4.170	3.164	2.558	2.153	1.866
180	E	30	3.604	2.341	1.703	1.327	1.083
		70	4.007	2.640	1.937	1.518	1.244
		140	4.390	2.914	2.148	1.688	1.387
		400	4.099	2.753	2.042	1.613	1.330
		750	4.017	2.713	2.019	1.598	1.320

TABLE 5.7.7
Lowlands Site Factor F_a from Malekmohammadi & Pezeshk, 2014

V_{S30}, m/s	Site Class	Depth, m	$S_S \le$ 0.25	$S_S = 0.50$	$S_S = 0.75$	$S_S = 1.00$	$S_S \ge 1.25$
560	C	30	2.156	1.844	1.622	1.457	1.330
		70	2.159	1.780	1.532	1.355	1.222
		140	2.068	1.653	1.398	1.221	1.091
		400	1.668	1.273	1.048	0.898	0.790
		750	1.283	0.953	0.772	0.654	0.570
270	D	30	2.182	1.586	1.242	1.022	0.870
		70	2.207	1.546	1.185	0.960	0.808
		140	2.131	1.448	1.090	0.872	0.727
		400	1.740	1.129	0.827	0.649	0.533
		750	1.348	0.851	0.614	0.476	0.388
180	E	30	2.072	1.351	0.987	0.773	0.636
		70	2.103	1.321	0.944	0.729	0.592
		140	2.035	1.241	0.871	0.663	0.534
		400	1.668	0.971	0.663	0.496	0.393
		750	1.296	0.743	0.493	0.365	0.286

TABLE 5.7.8
Lowlands Site Factor F_v from Malekmohammadi & Pezeshk, 2014

V_{S30}, m/s	Site Class	Depth, m	$S_1 \le$ 0.10	$S_1 = 0.20$	$S_1 = 0.30$	$S_1 = 0.40$	$S_1 \ge 0.50$
560	C	30	3.366	3.131	2.944	2.792	2.666
		70	4.494	4.175	3.921	3.716	3.546
		140	5.702	5.290	4.965	4.702	4.486
		400	4.215	3.918	3.682	3.490	3.332
		750	3.593	3.344	3.144	2.982	2.848
270	D	30	3.842	3.035	2.512	2.150	1.888
		70	4.449	3.509	2.901	2.482	2.178
		140	5.038	3.968	3.278	2.803	2.459
		400	4.443	3.507	2.901	2.482	2.179
		750	4.210	3.326	2.753	2.358	2.070
180	E	30	3.671	2.581	1.966	1.583	1.325
		70	4.067	2.854	2.173	1.748	1.462
		140	4.446	3.116	2.370	1.906	1.594
		400	4.142	2.909	2.216	1.783	1.492
		750	4.055	2.852	2.173	1.749	1.464

5.8   ASCE 43-05 AND ASCE 4-16 DESIGN GROUND MOTIONS

For nuclear, safety-related structures, ground motion definition is based on ASCE 43-05 (ASCE, 2005). Design ground motions in ASCE 43-05 are defined to be uniform hazard, not risk-targeted.

ASCE 43-05 defines Seismic Design Categories (SDCs) which are very different from SDCs defined in AASHTO and in ASCE 7-16. The mean recurrence interval (MRI) for design ground shaking in ASCE 43-05 is dependent upon SDC. H_D is the annual frequency of exceedance (equal to the inverse of MRI). A "design factor" (DF) is also incorporated into design ground motion determination in ASCE 43-05. Table 5.8.1 defines the parameters needed for design response spectra (DRS) calculation. Equations 5.8-1 through 5.8-4 summarize the calculations. The DRS is the design factor multiplied by the uniform hazard response spectrum (UHRS).

$$DRS = DF \cdot UHRS \tag{5.8-1}$$

$$DF = Max\{DF_1,\ DF_2\} \tag{5.8-2}$$

$$DF_2 = 0.60\left(A_R\right)^\alpha \tag{5.8-3}$$

$$A_R = \frac{SA_{0.1H_D}}{SA_{H_D}} \tag{5.8-4}$$

Given that mapped data are consistent with bedrock conditions, a procedure is necessary for translating bedrock motions to the surface. ASCE 43-05 specifies the following procedure:

(a) Convolve the bedrock spectral ordinates for $UHRS\text{-}H_D$ to the surface.
(b) Convolve the bedrock spectral ordinates for $UHRS\text{-}0.10H_D$ to the surface.
(c) Determine the slope factor, A_R, at each frequency.
(d) Determine DF at each frequency.
(e) Determine the DRS at each frequency.

TABLE 5.8.1
ASCE 43-05 Design Response Spectra (DRS) Parameters

SDC	H_D	MRI	DF_1	α	PGA_{min}
3	0.0004	2,500 years	0.8	0.40	0.06 g
4	0.0004	2,500 years	1.0	0.80	0.08 g
5	0.0001	10,000 years	1.0	0.80	0.10 g

The convolution of bedrock spectral ordinates to the surface may require site-specific, site response analysis techniques. For the examples presented here, these convolutions will be accomplished using code-based site factors.

Example 5.8-1: ASCE 43-05 Design Ground Motion

For example, consider a site in Oak Ridge, TN, with Class D Site Class assigned. For a Seismic Design Category 4 facility, Table 5.8.2 summarizes the spectral control points for H_D (2,500 year MRI) and for $0.1H_D$ (25,000 years *MRI*). Computation of the DRS in accordance with ASCE 43-05 is outlined below:

For A_S:

$$DF_1 = 1.0$$

$$DF_2 = 0.6\left(\frac{1.156}{0.424}\right)^{0.8} = 1.338$$

$$\left(A_S\right)_{DRS} = 1.338 \times 0.424 = 0.567$$

For S_{DS}:

$$DF_1 = 1.0$$

$$DF_2 = 0.6\left(\frac{1.528}{0.712}\right)^{0.8} = 1.105$$

$$\left(S_{DS}\right)_{DRS} = 1.105 \times 0.712 = 0.787$$

For S_{D1}:

$$DF_1 = 1.0$$

TABLE 5.8.2
UHRS Spectral Control Points – Oak Ridge, TN

RotD100 Spectra	MRI = $1/H_D$	
	2,500 year	25,000 year
PGA	0.3347	1.051
S_S	0.5120	1.528
S_1	0.1021	0.287
F_{PGA}	1.265	1.100
F_a	1.390	1.000
F_v	2.396	$1.5 \times 2.026 = 3.039$
UHRS-A_S	0.424	1.156
UHRS-S_{DS}	0.712	1.528
UHRS-S_{D1}	0.244	0.872

$$DF_2 = 0.6 \left(\frac{0.872}{0.244} \right)^{0.8} = 1.662$$

$$(S_{D1})_{DRS} = 1.662 \times 0.244 = 0.405$$

For SDC 5 facilities, note that the USGS Unified Hazard Tool does not include data and disaggregation for the 100,000-year *MRI* ground shaking. However, the USGS web-based services may be used to obtain data for this rare level of ground shaking.

For the Oak Ridge example outlined above, determine the 100,000-year *MRI* ground shaking parameters. The site coordinates are 36.01°, −84.27°. The USGS web services applications are not as user-friendly as is the Unified Hazard Tool, but the required data can be obtained, as shown in Figure 5.8.1. Other spectral values could be obtained in a similar fashion. The USGS web services application is useful for such analyses and may be accessed at usgs.gov/nshmp-haz-ws/apps/services.html . Note that changes in the specified URL are required to obtain the desired results.

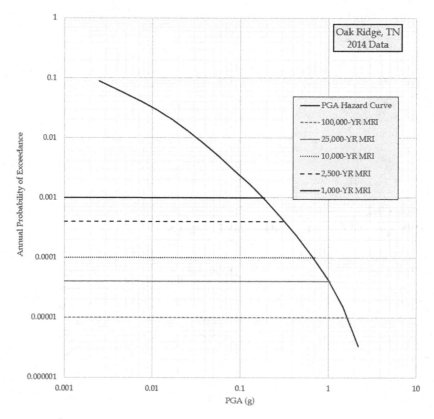

FIGURE 5.8.1   Oak Ridge, TN, hazard curve for *PGA*.

In addition to DRS computations, ASCE 43-05 is clear in defining requirements for ground motion histories to be used in structural analysis. The following requirements are included in ASCE 43-05:

(a) The frequency range of interest generally shall be 0.20–25 Hz.
(b) The largest permissible time increment for ground motion histories is 0.01 sec. If frequencies in excess of 50 Hz are of interest, then the maximum permissible time increment is decreased to $1/(2f)$, where f is the largest frequency of interest.
(c) The total duration shall be at least 20 sec.
(d) The 5% damped acceleration spectrum shall be computed and the suite average determined. The suite mean shall not fall below the target more than 10% at any one frequency, nor shall the suite mean exceed the target by more than 30% at any one frequency.
(e) Strong motion durations, measured by D_{5-75}, PGV/PGA, and $PGA(PGD)/(PGV)^2$, shall be generally consistent with the characteristic values for magnitude and distance for the site.
(f) Correlation of orthogonal components shall not exceed 0.30. This requirement is based on work found in the literature (Hadjian, 1981).

For safety-related nuclear structures, ASCE 4-16 (ASCE, 2017) primarily references ASCE 43-05 for development of design ground motions. ASCE 4 does provide the correlation equation for two components of ground motion, which is limited to no more than 0.30 for any single record pair, and no more than 0.16 for a suite average. σ_x and σ_y are standard deviation values:

$$\rho_{xy} = \left| \frac{1}{N} \sum_{i=1}^{N} \frac{(x_i - \bar{x})(y_i - \bar{y})}{\sigma_x \sigma_y} \right| \tag{5.8-5}$$

5.9   AASHTO DESIGN GROUND MOTIONS

AASHTO (AASHTO, 2011) design response spectra are uniform hazard, geometric mean based, and have 7% probability of exceedance in 75 years (*MRI* of about 1,000 years). AASHTO has its own site factors and they are currently different from those found in ASCE 7-16 for buildings. Tables 5.9.1, 5.9.2, and 5.9.3 give AASHTO site factors. Another difference between AASHTO and ASCE 7-16 is the peak ground acceleration at the surface, called A_S and equal to $F_{PGA} \times PGA$ in AASHTO. In ASCE 7-16, A_S is simply taken equal to $0.40 S_{DS}$. Other than these differences, the shape of the design response spectrum in the AASHTO specifications is developed in a similar fashion to that in ASCE 7-16.

However, as of the date of this discussion, projects are in progress that would significantly alter the nature of AASHTO design ground motions. The engineer is well served in keeping abreast of the nature of current design ground motions for all codes and specifications.

TABLE 5.9.1
AASHTO Site Factor F_{PGA}

Site Class	PGA Range of Applicability				
	0.10	0.20	0.30	0.40	0.50
A	0.80	0.80	0.80	0.80	0.80
B	1.00	1.00	1.00	1.00	1.00
C	1.20	1.20	1.10	1.00	1.00
D	1.60	1.40	1.20	1.10	1.00
E	2.50	1.70	1.20	0.90	0.90

TABLE 5.9.2
AASHTO Site Factor F_a

Site Class	S_S Range of Applicability				
	0.25	0.50	0.75	1.00	1.25
A	0.80	0.80	0.80	0.80	0.80
B	1.00	1.00	1.00	1.00	1.00
C	1.20	1.20	1.10	1.00	1.00
D	1.60	1.40	1.20	1.10	1.00
E	2.50	1.70	1.20	0.90	0.90

TABLE 5.9.3
AASHTO Site Factor F_v

Site Class	S_1 Range of Applicability				
	0.10	0.20	0.30	0.40	0.50
A	0.80	0.80	0.80	0.80	0.80
B	1.00	1.00	1.00	1.00	1.00
C	1.70	1.60	1.50	1.40	1.30
D	2.40	2.00	1.80	1.60	1.50
E	3.50	3.20	2.80	2.40	2.40

For requirements on ground motion selection, AASHTO is relatively silent. However, the FHWA retrofit manual for bridges (Buckle, Friedland, Martin, Nutt, & Power, 206) does contain a few requirements. These are summarized below:

- Either three ground motions or seven ground motions may be used in the analysis. If three ground motions are used, the maximum response of the three shall be used for design. If seven are used, the average response may be taken as the design value.

- The suite mean spectrum shall not fall below the target by more than 15% over the period range of interest, and the average ratio of suite mean to target over the entire period range of interest shall be at least 1.0.
- Amplitude scaled or spectrally matched records are permissible.
- For near-fault sites, ground motions are to be rotated to fault-normal (*FN*) and fault-parallel (*FP*) orientations for application in structural analysis.

5.10   GROUND MOTION SELECTION AND MODIFICATION

For structural design by response history analysis, it becomes necessary to select a suite of ground motion records. Typical suite sizes range from as few as 3 record pairs to as many as 11 or more. In fact, given the emphasis on performance-based design likely to occur in the near future, as many as 30 or 40 record pairs (or more) could be a necessary minimum in research when estimates of response variability to ground shaking are needed.

Ground motion selection requires careful attention to several factors. An excellent reference for the process is found in NIST-GCR-11-917-15 (NEHRP Consultants Joint Venture, 2011). Engineers and researchers working on ground motion selection would be well served in obtaining the freely available digital document.

Some of the parameters involved in selecting candidate ground motions for structural analysis include, in descending order of importance (author's opinion), at least for far-field, non-subduction earthquakes:

1. Match to spectral shape
2. Magnitude
3. Recording station site characterization
4. Distance
5. Fault type

Given that match to spectral shape is a critical factor in whether a particular ground motion should be considered for a given site, some means of measuring match to spectral shape is necessary. Two proposed measures of match to spectral shape are mean-square-error (*MSE*) and D_{RMS}. *MSE* and D_{RMS} are given in Equations 5.10-1 and 5.10-2. Notice that *MSE* has a scale factor, *f*. Equation 5.10-3 provides an expression for computing the scale factor, *f*, which minimizes *MSE* over a specified range of periods. This is the scale factor used in amplitude scaling of ground motion records. So *MSE* can be computed prescaling by inserting $f = 1$ in the equation for *MSE*, and post-scaling by inserting the computed scale factor for *f* in the equation for *MSE*. The post-scaled *MSE* is the appropriate value for assessing candidate records. The weights, $w(T_i)$, at each period are typically taken equal to 1 for all periods in the range of interest.

$$MSE = \frac{\sum w(T_i) \cdot \left\{ \ln\left[SA_{TARGET}(T_i) \right] - \ln\left[f \cdot SA_{GM}(T_i) \right] \right\}^2}{\sum w(T_i)} \qquad (5.10\text{-}1)$$

$$D_{RMS} = \frac{1}{N} \sqrt{\sum_{i=1}^{N} \left(\frac{(SA_{GM})_{Ti}}{PGA_{GM}} - \frac{(SA_{TAR})_{Ti}}{PGA_{TAR}} \right)^2} \qquad (5.10\text{-}2)$$

$$\ln f = \frac{\sum \left[w(T_i) \cdot \ln \frac{SA_{TARGET}(T_i)}{SA_{GM}(T_i)} \right]}{\sum w(T_i)} \qquad (5.10\text{-}3)$$

A knowledge of characteristic magnitude, distance combinations may be obtained for a given site by disaggregating the seismic hazard, available at the USGS online Unified Hazard Tool (UHT): usgs.gov/hazards/interactive/.

Example 5.10-1: Seismic Hazard Deaggregation

For a site near Memphis at coordinates 35°09'11"N, 90°03'07"W (+35.153, −90.052), deaggregation using the UHT reveals the results shown in Figures 5.10.1, 5.10.2, and 5.10.3. For all three spectral control points, the "modal" M, R combination is about $M_W7.55$, 55 km. The numerical results for S_1 are shown below. Those for PGA and S_s are similar. The site may be characterized as "uni-modal" since the modal M, R combination is constant across the control periods. The "modal" M, R combination is the one most likely, out of all contributions to the hazard at a site, to produce the design ground motion at any given period. For this

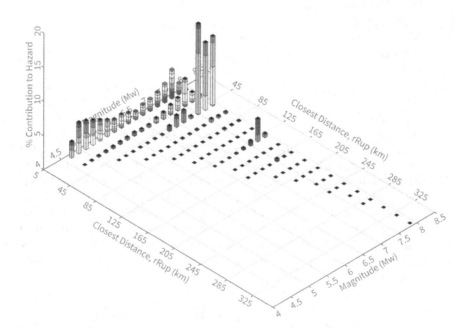

FIGURE 5.10.1 2,500-year *MRI PGA* deaggregation for Memphis, TN (USGS).

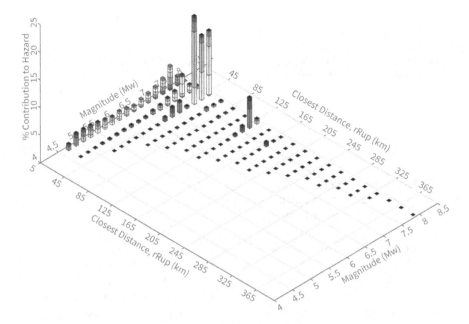

FIGURE 5.10.2 2,500-year *MRI S_S* deaggregation for Memphis, TN (USGS).

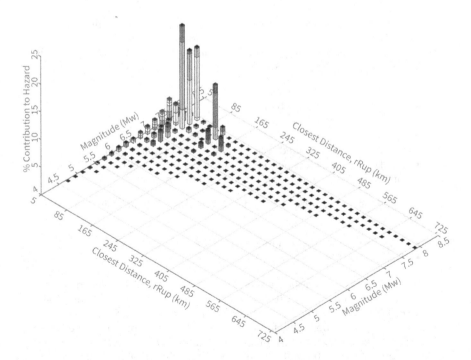

FIGURE 5.10.3 2,500-year *MRI S₁* deaggregation for Memphis, TN (USGS).

site near Memphis, ground motion records from large magnitude, far-field events would seem to be most appropriate.

Once a set of candidate records (usually 100 or more) has been established, a means of selecting the most appropriate and modifying the records is necessary. While many candidate records from a single event may be included, it is customary (and sometimes required) that no more than three or four records from any single event be included in the final suite.

There are at least three methods for modifying ground motion records for use in structural analysis:

1. Amplitude scaling
2. Spectral matching in the time domain
3. Spectral matching in the frequency domain

Each of these is discussed in detail.

Amplitude scaling is typically the preferred method for ground motion modification. The accelerations at each time step in the acceleration are all multiplied by some factor. The time scale is not adjusted in any fashion. Frequency content and pulse-type character of the ground motion are retained with amplitude scaling.

Example 5.10-2: PEER Ground Motion Selection and Modification

The PEER Ground Motion Database (Pacific Earthquake Engineering Research Center, 2014) includes an excellent online tool for ground motion selection with subsequent amplitude scaling. Consider a building at the site in Memphis discussed previously and having the following parameters defined:

- Modal $M,R = M_W 7.55$, 55 km
- S_{MS} = 1.235 g ($RotD100$-based)
- S_{M1} = 1.021 g ($RotD100$-based)
- T_L = 12 sec
- Structure period required to achieve 90% mass participation = 0.2 sec
- Fundamental structure periods = 1.5 sec (E-W), 1.8 sec (N-S)
- Site Class D (V_{S30} in the range of 180–360 m/sec)

Set a period range of interest in accordance with ASCE 7-16 provisions previously discussed of:

- T_{LOWER} = 0.20 sec, but not more than 0.2 × 1.5 = 0.30 sec
- T_{UPPER} = 2 × 1.80 = 3.6 sec

After defining the design response spectrum at the PEER GMDB, selection criteria were set as shown in Figure 5.10.4, resulting in 73 candidate record pairs. These 73 record pairs were sorted in ascending order according to *MSE*. The lower the post-scaled *MSE*, the better the fit to the target. The record pairs with the lowest *MSE* were selected since no event had more than 4 record pairs in the 11 lowest *MSE* records. Table 5.10.1 summarizes metadata for the final suite of 11 record pairs for use in structural analysis and design.

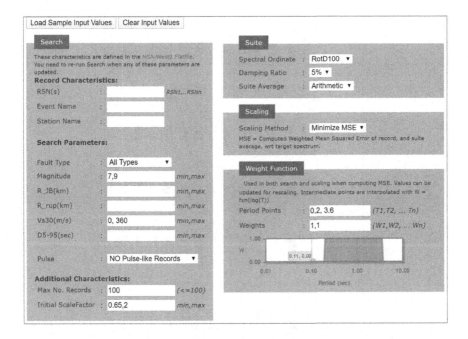

FIGURE 5.10.4 PEER ground motion database selection criteria.

Figures 5.10.5, 5.10.6, and 5.10.7 present the spectra plots, log-based standard deviation plot, and the suite mean-to-target ratios across the range of periods of interest, 0.20–3.60 sec. This initially proposed 11-record suite of ground motions is likely unacceptable for two or three reasons.

(a) The target variability, in terms of log-based standard deviation, is less than 0.60. If variability in response is of concern, then a value of 0.6, or a value from an appropriate ground motion model, should be targeted.

(b) The suite mean falls below 90% of the target within the period range of interest. This could be mitigated by amplifying all initially computed scale factors by the same value to achieve the target minimum of 90%.

(c) The average ratio of suite mean-to-target ratio within the period range of interest is less than 1.0. This could be mitigated by amplifying all initially computed scaled factors by the same value to achieve the target average of 1.0.

In addition to the online ground motion scaling tool available at the PEER Ground Motion Database (ngawest2.berkeley.edu), software available for ground motion scaling includes:

- *GMScaling-GeoMean-2020*.xlsx (author Huff)
- *SigmaSpectra* (Kottke & Rathje, 2012)
- *SeismoSelect* (SeismoSoft, 2020)

TABLE 5.10.1

Record Selection Details for Example Site

PEER RSN	MSE	f	D_{5-95} Sec	EQ Name	Year	M_W	R_{jb} km	V_{S30} m/sec	LUF Hz
1158	0.0642	1.232	12	Kocaeli	1999	7.51	14	282	0.100
1203	0.0409	1.691	33	Chi-Chi	1999	7.62	16	233	0.063
1495	0.0480	1.787	27	Chi-Chi	1999	7.62	6	359	0.250
1605	0.0395	0.987	11	Duzce	1999	7.14	0	282	0.100
5825	0.0761	1.475	44	El Mayor	2010	7.20	9	242	0.050
5827	0.0688	1.269	35	El Mayor	2010	7.20	13	242	0.063
5975	0.0422	1.771	42	El Mayor	2010	7.20	19	231	0.038
6890	0.0559	1.784	20	Darfield	2010	7.00	18	204	0.075
6923	0.1371	1.797	20	Darfield	2010	7.00	31	255	0.200
6952	0.0990	1.686	37	Darfield	2010	7.00	19	263	0.063
6953	0.0715	1.859	22	Darfield	2010	7.00	25	206	0.088

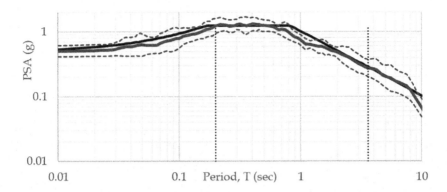

FIGURE 5.10.5   Target and suite mean *PSA*.

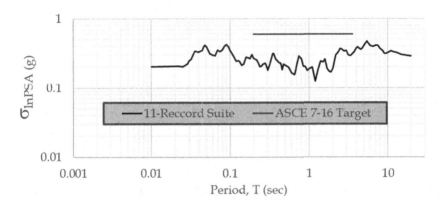

FIGURE 5.10.6   Target and suite variability.

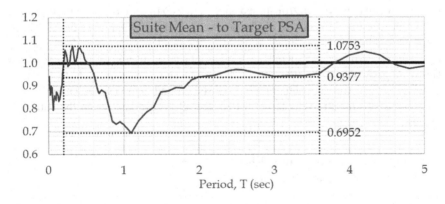

FIGURE 5.10.7   Ratio of suite mean-to-target *PSA*.

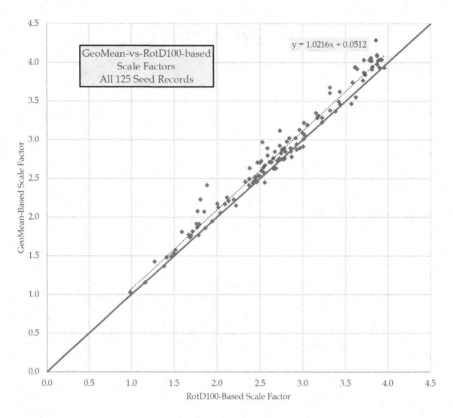

FIGURE 5.10.8   *RotD100* versus *GeoMean* scale factor computation.

GMScaling-GeoMean-2020 requires that the target spectrum be geometric mean based. Conversion factors identified in the section "Ground Motion Directionality" may be used to convert *RotD100*-based spectra to *GeoMean*-based spectra if needed. Such a procedure is illustrated in the literature (Huff, 2020). Figure 5.10.8 depicts sample scale factor computations, illustrating the fact that either *GeoMean*-based or *RotD100*-based target spectra may be used as long as the user knows the target basis. *GMScaling-GeoMean-2020* automatically produces informative plots to assist the engineer in determining whether their proposed ground motion suites satisfy criteria for closeness of match to the target. The Excel file is available from the author on request.

Example 5.10-3: Excel Spreadsheet Ground Motion Selection and Modification

Consider the problem just solved above using the online PEER ground motion scaling application. Select and scale a ground motion suite for the same conditions using *GM-Scaling-GeoMean-2020*.xlsx. Since *GM-Scaling-GeoMean-2020*.xlsx requires the target to be *GeoMean*-based, first convert the *RotD100*-based target used above to a *GeoMean*-based target. Recall that the *RotD100*-to-*GeoMean* ratio is 1.1 for S_S and 1.3 for S_1.

- S_{MS} = 1.235 g (*RotD100*-based)
- S_{MS} = 1.235 / 1.1 = 1.123 g (*GeoMean*-based)
- S_{M1} = 1.021 g (*RotD100*-based)
- S_{M1} = 1.021 / 1.3 = 0.785 g (*GeoMean*-based)

Table 5.10.2 summarizes metadata for results from *GM-Scaling-GeoMean-2020*. xlsx. Figures 5.10.9 and 5.10.10 depict the spectra plots and suite mean-to-target plots for the records selected using *GM-Scaling-GeoMean-2020*.xlsx.

Note that the minimum suite mean-to-target ratio within the period range of interest is 0.895. Thus, all scale factors would need to be adjusted by a factor of 0.900 / 0.895 = 1.006 in order to satisfy criteria that this parameter be no less than 0.900. A plot of log-based standard deviation would reveal results similar to that for the 11-record suite generated using the PEER ground motion database – namely, values well below 0.60. If variability in response is of no concern, this suite, with scale factors adjusted as noted above, could be a valid suite of records. If variability in response is needed, then a suite possessing a higher log-based standard deviation will be necessary.

SigmaSpectra (github.com/arkottke/sigmaspectra) offers the advantage of permitting the user to specify a target log-based standard deviation. ASCE 7-16 suggests a log-based standard deviation equal to 0.60 across all periods of interest. An appropriate ground motion model could also be used to establish a target log-based standard deviation. This preserves record-to-record variability with a small subset of all available ground motions. The procedure could potentially be used to design for responses greater than the median (by some specific number of standard deviations).

TABLE 5.10.2
Alternative Scaling Metadata

RSN	EQ	Year	M_W	Site Class	f
6959	Darfield	2010	7.00	E	1.75
8606	El Mayor	2010	7.20	D	1.91
1187	Chi-Chi	1999	7.62	D	2.63
3680	SMART1(45)	1986	7.30	D	2.86
Non-Peer 426	Ecuador	2016	7.80	E	1.56
1481	Chi-Chi	1999	762	D	2.76
5823	El Mayor	2010	7.20	D	2.38
5988	El Mayor	2010	7.20	D	2.62
1158	Kocaeli	1999	7.51	D	1.53
577	SMART1(45)	1986	7.30	D	2.91
1605	Duzce	1999	7.14	D	1.17

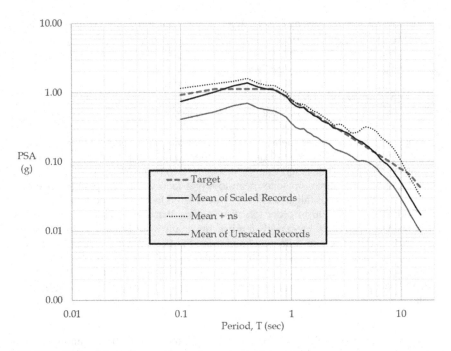

FIGURE 5.10.9   Alternative scaling suite mean and target *PSA*.

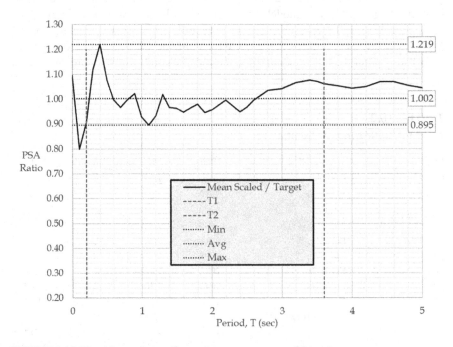

FIGURE 5.10.10   Alternative scaling suite mean-to-target *PSA* ratio.

Example 5.10-4: *SigmaSpectra* Ground Motion Selection and Modification

SigmaSpectra may be used with the same 11-record suite obtained from *GM-Scaling-GeoMean-2020* to obtain scale factors that will produce both a close fit to the target *PSA* and a close fit to the target $\sigma_{\ln PSA}$. The *GeoMean*-based target spectrum may be pasted from Excel into *SigmaSpectra*. The folder containing the records may then be specified, as shown in Figure 5.10.11. Figure 5.10.12 is a plot of the results from SigmaSpectra scaling. The log-based standard deviation has an acceptable match to the target value of 0.60. However, the minimum suite-to-target *PSA* ratio is less than 0.900, as indicated in Figure 5.10-12. The scale factors reported in Table 5.10.3 would need to be amplified by a factor equal to 0.900/0.876 = 1.027 to satisfy the requirement that the minimum suite-to-target *PSA* be no less than 0.900 across the period range of interest.

SeismoSelect provides another alternative for ground motion scaling. Included in the software are various options for target basis (*RotD100*, *GeoMean*, etc.), as well as for ground motion database sources (PEER, ESMD, etc.). *SeismoSelect* also has features that enable the user to generate code-based target spectra for many different specifications.

SeismoMatch (SeismoSoft, 2020) adds wavelets to an accelerogram to create a new accelerogram whose *PSA* response spectrum matches, as closely as possible, the target spectrum. Modification is performed directly on the accelerograms in the

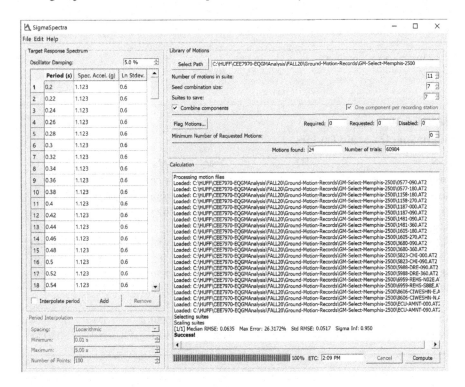

FIGURE 5.10.11 SigmaSpectra ground motion scaling.

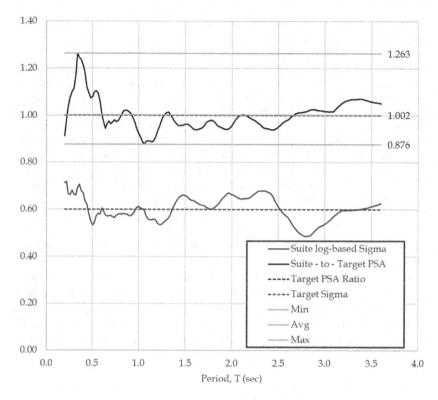

FIGURE 5.10.12   SigmaSpectra suite plots.

TABLE 5.10.3
SigmaSpectra Scale Factors

RSN	EQ	Year	M_w	Site Class	f
6959	Darfield	2010	7.00	E	2.26
8606	El Mayor	2010	7.20	D	2.16
1187	Chi-Chi	1999	7.62	D	2.28
3680	SMART1(45)	1986	7.30	D	1.50
Non-Peer 426	Ecuador	2016	7.80	E	2.41
1481	Chi-Chi	1999	762	D	2.03
5823	El Mayor	2010	7.20	D	2.41
5988	El Mayor	2010	7.20	D	1.75
1158	Kocaeli	1999	7.51	D	2.90
577	SMART1(45)	1986	7.30	D	1.04
1605	Duzce	1999	7.14	D	3.31

time domain. Spectral matching, whether time domain based or frequency domain based, is very different from amplitude scaling and may dramatically alter the basic character of the matched record.

Example 5.10-5: Spectral Matching in the Time Domain

Consider the Kocaeli record pair (RSN 1158) selected for the 11-record suite using *GM-Scaling-GeoMean.xlsx* in Example 5.10-3. The computed scale factor to minimize *MSE* in the period range of 0.20–3.60 sec was 1.526. Load each of the two components into *SeismoMatch* and apply the scale factor. Further, define the target spectrum used in that section in *SeismoMatch* by specifying a file to be read in defining the target. Specify the same period range for matching as was used in scaling and perform the matching process to obtain two new accelerograms.

Care must be taken in spectral matching to ensure that "realistic" (somewhat subjective) ground motions are produced in the matching process. Figure 5.10.13 shows the pre-matched spectrum for record RSN 1158. Figure 5.10.14 is the corresponding post-matched spectra, revealing a very close match between the record spectra and the target over the period range of interest. Figures 5.10.15 and 5.10.16 depict the pre- and post-matched displacement histories for the two components of RSN 1158. Table 5.10.4 is a summary of pre- and post-matched ground motion parameters. As long as pulse character is of no concern, and as long as record-to-record variability is of no concern, the matching process appears to have produced "realistic" records.

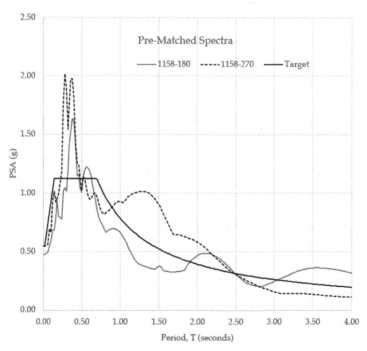

FIGURE 5.10.13 Pre-matched ground motion spectra.

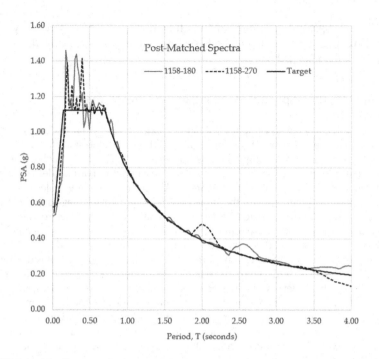

FIGURE 5.10.14   Post-matched ground motion spectra.

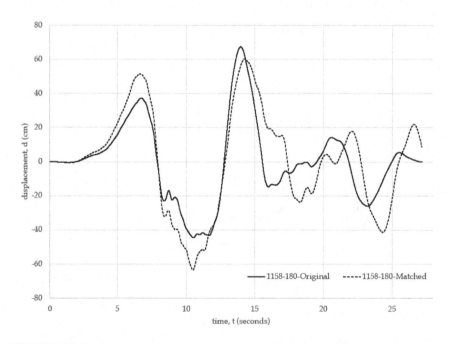

FIGURE 5.10.15   Pre- and post-matched displacement histories – 1158 NS component.

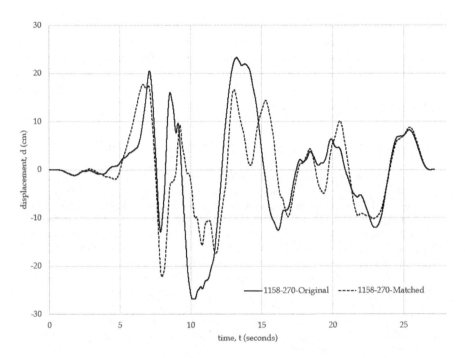

FIGURE 5.10.16   Pre- and post-matched displacement histories – 1158 EW component.

TABLE 5.10.4
Pre- and Post-Matched Ground Motion Parameters

	Pre-Matched		Post-Matched	
Accelerogram	1158-180	1158-270	1158-180	1158-270
Max. Acc. (*g*)	0.476	0.546	0.522	0.544
Max. velocity (cm/s)	89.8	70.8	94.0	69.3
Max. displacement (cm)	67.3	26.9	65.5	21.9
Arias intensity	2.53	3.09	3.95	3.22
Cumulative absolute velocity	1,296	1,210	1,789	1,389

SeismoArtif (SeismoSoft, 2020) has multiple capabilities, among which is spectral matching in the frequency domain. The Fourier spectrum for a record is first computed and compared to a Fourier spectrum generated from the target *PSA* spectrum. Adjustment to the ground motion Fourier spectrum is made to produce a closer match to the target Fourier spectrum. The modified Fourier spectrum is then converted back to a new accelerogram.

Example 5.10-6: Spectral Matching in the Frequency Domain

As an example, consider RSN 1158-180 from Example 5.10-5. Load the target spectrum into *SeismoArtif* and select the option to "adjust a real accelerogram." Figure 5.10.17 shows the displacement history which results from the attempted matching in the frequency domain. The resulting record is dramatically different from the original, as are the ground motion parameters. An end-of-record drift of about 30 cm exists. So, it becomes clear that care should be exercised in frequency domain matching of real records. This frequency-domain-matched record would not be a viable candidate for inclusion in a final suite of records to be used in a structural design.

Physics-based models exist for generating synthetic ground motions. *SeismoArtif* has this capability, in addition to those previously mentioned. Near-fault and far-field options are available. Site conditions and tectonic environment (intraplate vs. plate boundary) are specified by the user.

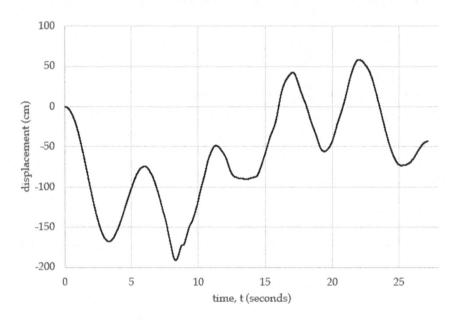

FIGURE 5.10.17   Frequency-domain-matched displacement history – RSN 1158-180.

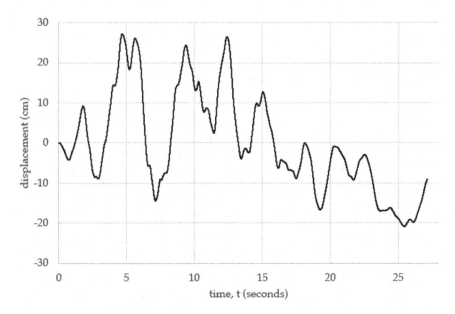

FIGURE 5.10.18   Synthetic adjusted accelerogram.

Example 5.10-7: Synthetic Record Generation and Adjustment

Using the synthetic accelerograms generation and adjustment, define the same target as used in the previous section on frequency domain matching of real records. Specifying an intraplate, $M_W7.7$, Site Class D condition, *SeismoArtif* produces eight synthetic records, which may be adjusted to the target. Figure 5.10.18 represents one of the eight resulting displacement histories. While a reasonably close match to the target spectrum is possible, the ground motion appears relatively "unrealistic." This should not be interpreted as an argument against synthetic or artificial accelerograms. However, such records should not be used blindly, without careful examination, and many iterations may need to be completed in order to generate acceptable records.

Artificial records are generated from an initial random noise signal, constrained by one of several available envelope functions, and adjusted in the frequency domain in *SeismoArtif*. For the target spectrum defined in the previous section and for a Saragoni & Hart envelope function (90 sec defined duration, 20 sec ramp-up time, t_1, and end intensity, $I_{dur} = 0.01$), the resulting displacement histories were found to be similar in nature to those produced from previous frequency domain matching procedures – "unrealistic" in character.

A simple model for artificial record generation is available through the State University of New York (SUNY) program *RSCTH*. Required input is minimal, including moment magnitude, epicentral distance, target spectrum parameters, and tectonic regime (Eastern or Western United States). The program is DOS-based.

Example 5.10-8: Artificial Ground Motion Development

A sample *RSCTH* input file is shown in Figure 5.10.19. The resulting artificial record, after baseline correction, is shown in Figure 5.10.20. A comparison between the specified "NEHRP-based" target spectrum and the ground motion spectrum is depicted in Figure 5.10.21. The "NEHRP-based" target spectrum uses 200 log-spaced points for target spectrum calculation based on S_{DS}, S_{D1}, T_S, and T_o input parameters. The resulting ground motion spectrum will typically have isolated periods for matching to the target at longer periods, as is evident in Figure 5.10.21.

A user-defined target spectrum often provides better results, at least with regard to match to the target. The same target spectrum as used for the "NEHRP-based" spectrum defined by 99 points in a user-defined target produces the ground motion history shown in Figure 5.10.22 and the spectra in Figure 5.10.23. Again, while match to spectral shape for artificial ground motions may appear impressive, such records should be used with caution in structural analysis.

```
 rscth.inp - Notepad

File Edit Format View Help
USER COMMENTS (min. 1 and max. 80 char.)
   EQGMA Course Example (SDS = 0.95 g, SD1 = 0.65 g)
IREGION      ISLIP      V30
   1            3       760.0
RMW     EPDIS     DAMP     IDURPM     NT       DT
 7.7     55      0.05       1      18000     0.005
NEHRP    T0       Sds       Ts       Sd1      Nper
   1    0.137    931.0     0.684    638.0      200
ITRSTP    ITRSFCN   NTRS   NSPEC
   0         2       5      128
  0.07      10.0
  0.63      50.0
  2.40     300.0
  7.16     200.0
 20.37      20.0
```

FIGURE 5.10.19 RSCTH sample input file.

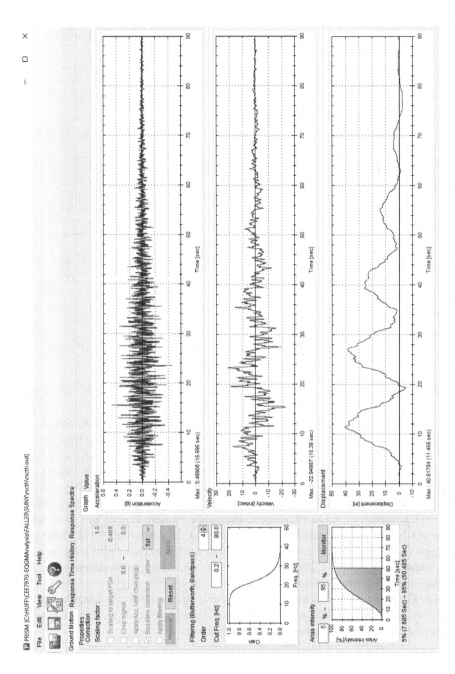

FIGURE 5.10.20 RSCTH artificial ground motion record.

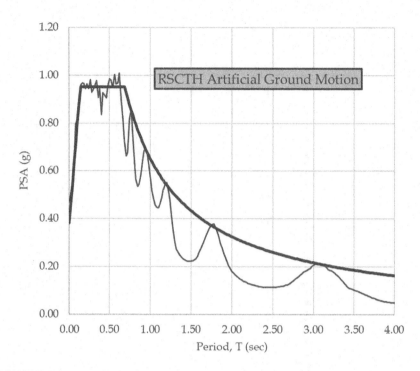

FIGURE 5.10.21   RSCTH artificial ground motion record spectrum versus target.

Target spectra for the examples considered here have been either *GeoMean*-based, uniform hazard (UHRS), or *RotD100*-based, risk-targeted (RTRS). Much research has been done on the nature of appropriate target spectra for ground motion selection and modification. The UHRS is generally accepted to be a conservative target, presumably the RTRS as well. Some would argue that these are overly conservative and that a conditional mean spectrum (CMS) is a more appropriate target. Rationale for the argument that the UHRS and RTRS are overly conservative is related to the fact that these spectra do not generally represent a single earthquake scenario, but an envelope of multiple scenarios. Figure 5.10.24 shows a type of CMS target spectrum in which the target is derived from a UHRS based on a single conditioning period, 1.0 sec. The spectrum has been derived from work in the literature (Kishida T., 2017) which permits either a single conditioning period or multiple conditioning periods.

The CMS and its variations will likely receive much attention in future research. For now, either the UHRS or the RTRS used for design seems to be a reasonable target for ground motion selection and modification.

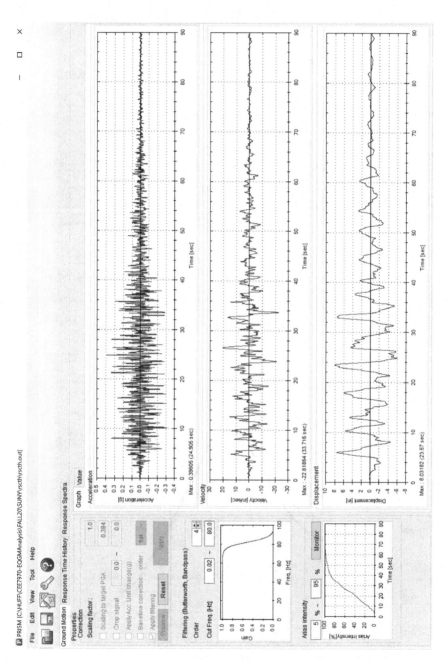

FIGURE 5.10.22 RSCTH artificial record with user-define target.

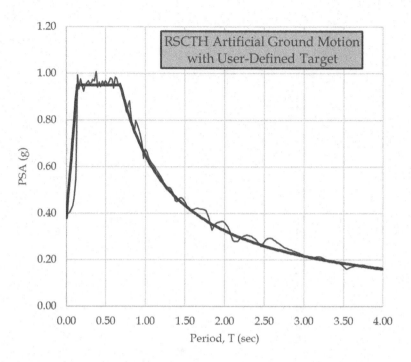

FIGURE 5.10.23   RSCTH artificial record spectra with user-defined target.

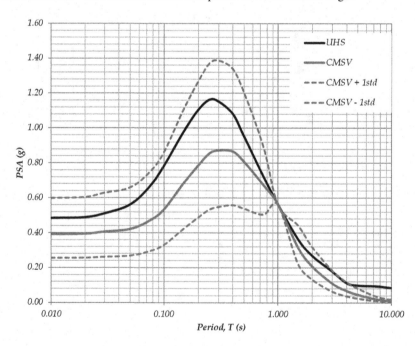

FIGURE 5.10.24   Conditional mean spectrum example.

5.11   GROUND MOTION DATABASES

Before detailed discussion of ground motion analysis, identification of sources for ground motion records is appropriate. Reliable ground motion records in terms of consistent processing of records, suitability for analysis of ground motion parameters, and suitability for use in structural analysis are available from several sources.

The PEER NGA-West2 database (Pacific Earthquake Engineering Research Center, 2014) is likely the first choice for the engineer needing ground motion records for structural analysis or ground motion analysis. PEER records have been consistently corrected and filtered according to the latest science on the subject, and are consistent in format.

In addition to accelerograms, PEER provides "flatfiles" for the records available in the PEER Ground Motion Database. A flatfile provides critical information, frequently referred to as "metadata," for the recording station and the seismic event. Examples of information available in the flatfiles include:

- Peak ground acceleration
- Peak ground velocity
- Peak ground displacement
- V_{S30} at the station
- Source-to-site distance measures
- Pseudo-acceleration response spectra
- Lowest usable frequency of the recorded motion
- Earthquake name and coordinates
- Station name and coordinates
- Earthquake magnitude

While it might seem logical to access the NGA-East database (Pacific Earthquake Engineering Research Center, 2014) when searching for records appropriate for use in the Central and Eastern United States (CEUS), there are very few strong ground motions available in this database due to the relative rarity of large magnitude events in the region. The database is likely to grow in the future, possibly with synthetic or artificial large magnitude records.

The CESMD database (USGS, CGS, ANSS, 2014) is also a valuable resource for ground motion records. The CESMD database is growing and user-friendly tools are available at CESMD. Figure 5.11.1 shows an example of a station map for the 2016 $M_w7.8$ Amberley, New Zealand, earthquake, also known as the Kaikoura earthquake. Figure 5.11.2 shows some of the plots available at CESMD. Ground motion plots similar to Figure 5.11.3 provide a valuable tool available at CESMD.

The COSMOS (USGS, CGS, ANSS, 2014) Virtual Data Center (VDC) is managed by the same group, as is CESMD. However, COSMOS does have many records that are not located at CESMD. These records are more likely to require baseline adjustment and filtering, which are discussed in another section.

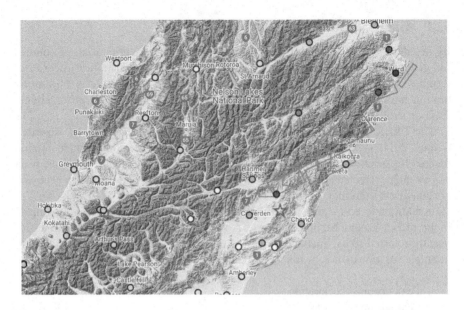

FIGURE 5.11.1   CESMD station map for 2016 Amberley, New Zealand, Earthquake.

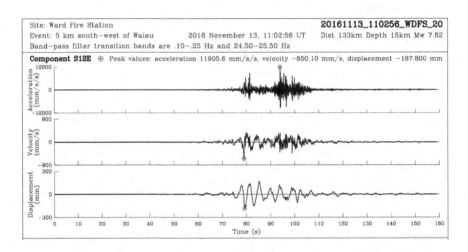

FIGURE 5.11.2   Amberley New Zealand Earthquake – Ward Fire Station – CESMD.

The European Strong Motion Database (Luzi, Puglia, & Russo, 2016) has a large collection of ground motion records with associated data. Some events, such as the 1975 Romania earthquake, have records available only here.

Records from New Zealand earthquakes are available at GeoNet (Earthquake Commission and GNS, 2020). Flatfiles which contain metadata (site class, source-to-site distance, PGA, PSA, etc.) are also provided. Processed records deemed appropriate for structural analysis have been identified and made available.

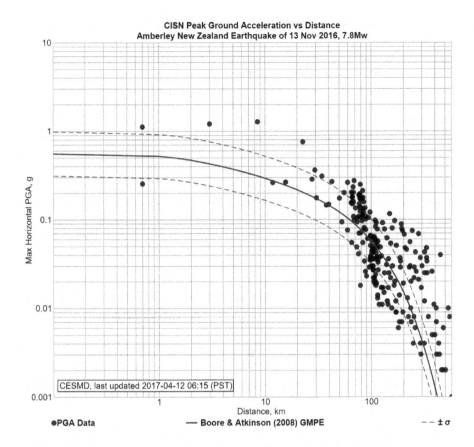

FIGURE 5.11.3   Amberley, New Zealand, Ground Motion Plot from CESMD.

The K-Net/Kik-Net database makes available records for events that have occurred in Japan (Network Center for Earthquake, Tsunami and Volcano, NIED, 2020).

For subduction zones, the UCLA B. John Garrick Institute for the Risk Sciences currently provides an initial set of 500 record pairs from subduction events. The NGA-Sub project will presumably include tens of thousands of records from multiple subduction events.

When ground motions are recorded, "noise" is present in the record. This "noise" must be removed prior to using the record for estimating ground motion parameters, such as peak ground velocity (PGV) and peak ground displacement (PGD).

5.12   GROUND MOTION BASELINE ADJUSTMENT AND FILTERING

Ground motions are recorded as three perpendicular components: two horizontal and one vertical. Therefore, the ground motion recorded is dependent upon the orientation of the instrumentation. Many times (but certainly not always) instrumentation has been oriented so as to register North-South and East-West components of ground motion.

When a ground motion is recorded during an earthquake, the resulting signal possesses "noise." The accelerograms really needs to be (a) baseline corrected and (b) filtered appropriately prior to use in parametric analysis or structural analysis.

Baseline adjustment removes spurious baseline trends, evident through observation of the integrated displacement history, through least squares fit of a polynomial (up to degree 3 in SeismoSignal, degree 2 in PRISM) with subsequent correction in the recorded accelerogram.

Filtering involves the removal of certain frequencies from the recorded motion. Filter types include Butterworth, Chebyshev, and Bessel, and may be "low-pass," "high-pass," "band-pass," or "band-stop." A "low-pass" filter suppresses frequencies higher than a user-specified value. "High-pass" filters allow only those frequencies higher than a user-specified value. "Band-pass" filters allow only those frequencies within a user-specified frequency range, while "band-stop" filters suppress frequencies within a user-specified range.

Furthermore, filters may be "causal" or "acausal." "Acausal" filtering involves applying a causal filter twice, once forward and once backward, through the ground motion record.

Example 5.12-1: Ground Motion Baseline Adjustment and Filtering

An example of an uncorrected, unfiltered ground motion record is shown in Figure 5.12.1. The filtered record is shown in Figure 5.12.2. Both records were loaded with PRISM.

A few observations in examining the raw and adjusted (corrected and filtered) records will prove informative.

- It is difficult to distinguish between the two records, raw and adjusted, by examining the ground acceleration history alone.

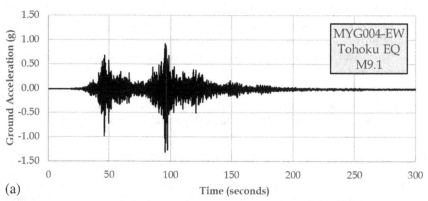

(a)

FIGURE 5.12.1 2011 M_w9.12 Japan Earthquake – MYG004 E-W (Raw). (a) Ground acceleration. (b) Ground velocity. (c) Ground displacement.

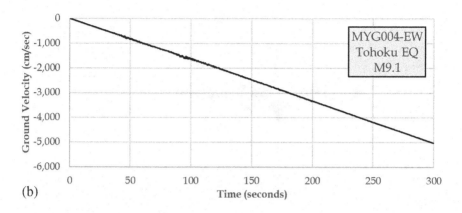

(b)

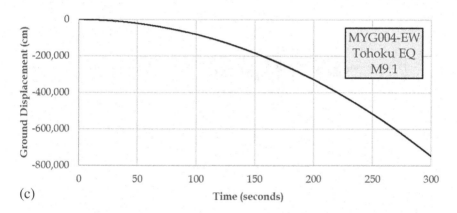

(c)

FIGURE 5.12.1 Continued.

- The ground acceleration values for the raw and adjusted records are similar.
- The ground velocity returns to zero for the adjusted record, as would be physically necessary. Unless a permanent ground displacement was observed in field reconnaissance, the same should be true of displacement.
- Integration of the acceleration history to obtain the ground velocity history and subsequent integration of the ground velocity history to obtain the ground displacement history are necessary to distinguish the raw and adjusted records.
- Estimating peak ground velocity (*PGV*) and peak ground displacement (*PGD*) requires careful attention to baseline adjustment and filtering.

Much attention has been devoted to the science of appropriately adjusting recorded accelerograms through baseline correction and filtering (Boore & Bommer, 2005) (Boore D. M., 2001) (Boore & Akkar, 2003).

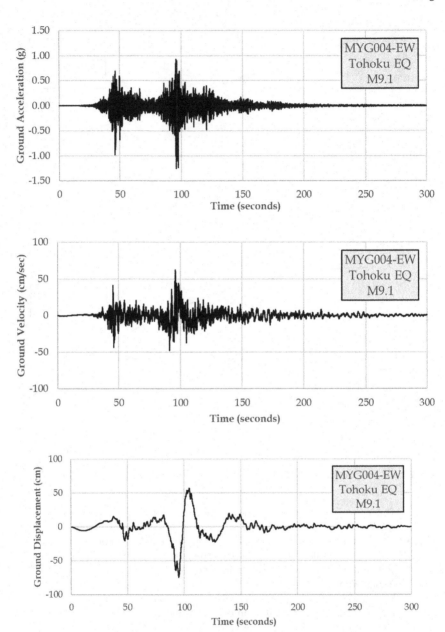

FIGURE 5.12.2 2011 M_w9.12 Japan Earthquake –MYG004 E-W (adjusted and filtered).

5.13   INCREMENTAL DYNAMIC ANALYSIS

For the seismic analysis and evaluation of structures, incremental dynamic analysis (IDA) has been proposed as a valuable tool, and justifiably so. In IDA, the structure is subjected to a suite of ground motions, each initially normalized in some fashion to a particular intensity measure (IM), at various levels of ground shaking by scaling each of the records progressively and determining some damage measure (DM).

One of the earlier treatments of IDA was published by Vamvatsikos and Cornell (August 2005). Intensity measures suggested in the study include peak ground acceleration, peak ground velocity, and spectral acceleration at a particular period, among others. Potential damage measures identified include base shear, element ductility, and interstory drift. At times, it may be advisable to adopt multiple damage measures for a structural evaluation.

A simple example will be useful in understanding the IDA concept.

Example 5.13-1: Incremental Dynamic Analysis

Table 5.13.1 lists the ground motion pairs scaled to a target response spectrum given by values for S_{DS} and S_{D1} of 0.643 and 0.473, respectively. The period range of interest is 0.20–3.00 sec, somewhat arbitrary given the single-degree-of-freedom system being analyzed. Figure 5.13.1 depicts the match to both the target PSA and the target log-based standard deviation of PSA, 0.60.

A simple SDOF structure is to be analyzed using IDA. Intensity is to be scaled to 20%, 40%, 60%, 80%, and 100% of the indicated scale factors in the table. Inelastic displacement is selected as the damage measure. Preliminary, approximate analyses indicate an estimated ductility demand of six. The structure properties are as follows:

- W = 1,000 kips
- Initial stiffness, k_i = 102 kips/inch
- Post-yield stiffness ratio α = 0.03
- Initial elastic damping = 1%
- Yield strength, F_y = 123 kips

Figure 5.13.2 is one example of information obtained from IDA. For this simple example, the normalization procedure is scaling to the intensity measure, spectral shape, and intensity over the period range T = [0.20–3.00] sec. The damage measure is structural displacement. Note that the yield displacement and target displacement may be easily calculated in this case as follows, and are shown in Figure 5.13.2.

$$\Delta_y = \frac{123}{102} = 1.206 \text{ inch}$$

$$\Delta_{Max} = \mu\Delta_y = 6 \cdot 1.206 = 7.235 \text{ inch}$$

TABLE 5.13.1

30-Record Ground Motion Suite for IDA

Earthquake Name	PEER RSN	f
1952 Kern Co.	12	4.096
1986 Taiwan SMART1 (45)	577	2.563
1986 Taiwan SMART1 (45)	582	2.279
1989 Loma Prieta	776	3.485
1989 Loma Prieta	786	2.505
1992 Landers	832	2.752
1992 Landers	859	3.896
1992 Landers	862	2.816
1992 Landers	864	2.629
1992 Landers	871	4.860
1992 Landers	890	3.849
1994 Northridge	1014	4.592
1994 Northridge	1025	4.071
1999 Chi-Chi, Taiwan	1198	2.563
1999 Chi-Chi, Taiwan	1203	2.953
1999 Chi-Chi, Taiwan	1206	3.466
1999 Chi-Chi, Taiwan	1208	2.500
1999 Chi-Chi, Taiwan	1214	4.453
1999 Chi-Chi, Taiwan	1227	2.355
1999 Chi-Chi, Taiwan	1228	3.223
1999 Chi-Chi, Taiwan	1245	4.440
1999 Chi-Chi, Taiwan	1277	3.028
1999 Chi-Chi, Taiwan	1351	2.845
1999 Chi-Chi, Taiwan	1375	4.180
1999 Chi-Chi, Taiwan	1579	4.362
2007 Chuetsu-oki	5284	2.936
1999 Hector Mine	1799	3.482
1999 Hector Mine	1818	3.849
1992 Landers	3758	2.977
1994 Northridge	1057	2.914

In Figure 5.13.2, the assumption has been made that the target spectrum is geometric mean based. Scaling was performed under this assumption. For each of the 30 record pairs, uncoupled analysis for each axis was performed, the maxima in each direction obtained, and the geometric mean of the maxima computed. If the target spectrum basis were maximum direction, then the procedure would be to determine the resultant displacement at each time step for each record, and then take the maximum resultant of the two perpendicular component responses. This is not the same as taking the resultant of the maxima in each direction, which would be erroneous since the maxima in each direction do not generally occur simultaneously.

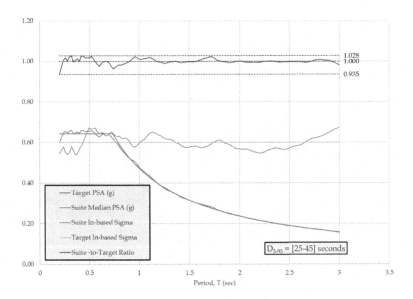

FIGURE 5.13.1   Match to target spectra for IDA.

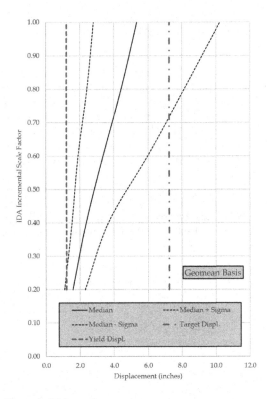

FIGURE 5.13.2   Example IDA results.

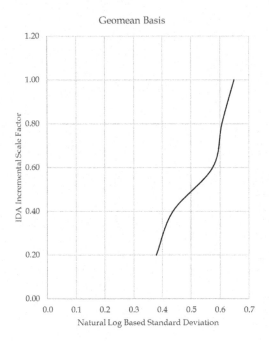

FIGURE 5.13.3   Variability in example IDA results.

From Figure 5.13.2, it may be inferred that the structure would be expected to yield at very low intensity levels. The median displacement is greater than the yield displacement, even at an intensity equal to 20% of the design intensity. At about 72% of the design intensity, the "Median plus Sigma" displacement is less than the target displacement. At full design intensity, while the "Median" displacement is less than the target displacement, the "Median + Sigma" displacement exceeds the target. The "Median" has been taken as representative of the geometric mean in this context.

Figure 5.13.3 depicts the change in natural log-based standard deviation of response as the IDA scaling is increased from 0.2 to 1.0. Recalling that the target is 0.60 for elastic *PSA*, it may be observed that at an intensity of about 75% of the design intensity (IDA Scale Factor = 1.0), the variability in inelastic displacement exceeds the variability in *PSA*.

6 Problems for Solution

PROBLEM 2.1

The W21x93 beam-column shown is oriented for strong-axis bending, is made from A913 Grade 65 steel, and is braced only at the ends. Does AISC 360-16 Appendix 1 permit design by inelastic analysis based on the given loads and geometry? M_2 acts in the direction shown. The direction of M_1 is to be determined from the given loads and geometry. Show all work.

- $P_u = 344$ kips
- $V_u = 171$ kips

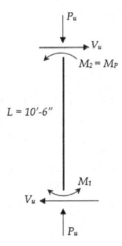

PROBLEM 2.2

The W21x93 beam-column shown is oriented for strong-axis bending, is made from A913 Grade 65 steel, and is braced only at the ends. Does AISC 360-16 Appendix 1 permit design by inelastic analysis based on the given loads and geometry? M_2 acts in the direction shown. The direction of M_1 is to be determined from the given loads and geometry. Show all work.

- $P_u = 344$ kips
- $V_u = 57$ kips

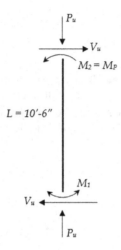

PROBLEM 2.3

The W24x84 beam shown is made from A992 steel and is braced only at the ends and at the load. The left end is fixed and the right end is a roller support.

- Using inelastic analysis and ignoring the beam weight, determine the maximum load, P_u, that can be supported.
- Establish whether the inelastic analysis is permissible by AISC 360-16 for this beam with the bracing scheme given.

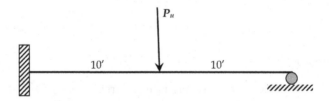

PROBLEM 2.4

The W12x65 beam shown is made from A992 steel and is braced only at the ends and at the load. The left end is fixed and the right end is a roller.

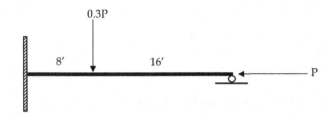

(a) Using inelastic analysis, ignoring the effect of the axial load, and ignoring the beam weight, determine the maximum load, P, that can be supported.
(b) Using inelastic analysis, including the effect of the axial load, and ignoring the beam weight, determine the maximum load, P, that can be supported. Use the P-M relationship used in class: $M_P' = 1.18 M_P (1 - P/P_y) \le M_P$
(c) Determine if the flange is ductile-compact according to AISC 360-16, Appendix 1 criteria.
(d) Determine if the web is ductile-compact according to AISC 360-16, Appendix 1 criteria.
(e) Determine the ductility-limited unbraced length, L_{pd}, for the 8-ft-long segment.
(f) Determine the ductility-limited unbraced length, L_{pd}, for the 16-ft-long segment.
(g) Is design by inelastic analysis permitted for this beam by AISC 360-16, Appendix 1? Why or why not?

PROBLEM 2.5

A Risk Category II structure in Cookeville, TN, is to be designed for a displacement ductility equal to no more than 6.0. The hysteretic behavior is to be modeled as bilinear with a postyield stiffness ratio of 2%. The inherent, elastic component of damping is taken to be 1% of critical. The hysteretic damping for nonlinear RSA is to be taken as 25% of that given by the theoretical equation for bilinear behavior. The EC8 pulse-type ground motion rule for response modification due to increased damping is to be used. The structure weight is 2,375 kips. The initial stiffness is 169 kips/in. The Site Class is "C."

(a) Determine S_{DS} and S_{D1} using ATC Hazard by Location.
(b) Determine the required strength, F_y.
(c) Determine the yield displacement, Δ_y, and the maximum displacement, Δ_{Max}.
(d) Determine the displacement amplification factor, $C\mu$.

PROBLEM 2.6

The lateral force-deformation relation of a SDOF system is idealized as elastic-perfectly plastic. In the linear elastic range of vibration, this SDOF system has the following properties:

- $k = 2.112$ kips/in
- $\xi = 2\%$
- $f_y = 5.55$ kips
- $W = 5,200$ lb

(a) Determine the natural period and damping ratio of this system vibrating in the elastic range.
(b) Determine the displacement ductility, μ, and inelastic amplification, $C\mu$, of the system subjected to the N/S Component of the El Centro ground motion scaled by a factor of three.
(c) Using the ductility calculated in part (b), run an effective linear analysis based on secant stiffness and effective damping ξ_{eff}.
(d) Compare the displacement from the nonlinear analysis in part (b) to that from the equivalent linear analysis in part (c). Determine the effective damping that would produce the correct displacement.

$$\xi_{eff} = \xi_o + \frac{2(\mu - 1)(1 - \alpha)}{(1 + \alpha\,\mu - \alpha)}$$

PROBLEM 3.1

For the portal frame shown, assess the columns and the beam for the load combination:

$$U = 1.2D + 1.0L + 1.0W$$

- Use a first-order analysis
- Use an AISC Direct Analysis Method solution

A913 Grade 70 steel is used for all members. Column bases are pinned. Vertical loads not explicitly included in the model shown (vertical loads on "leaning columns") include 600 kips dead load and 240 kips live load.

Vertical loads are 300 kips dead load and 120 kips live load at the top of each column. The lateral load due to wind is 75 kips. The frame is braced out of plane.

The lateral drift, from a first-order analysis, of the frame due to the 75 kips wind load is 3.47 inches. This drift was computed using nominal stiffness values, EA and EI (no stiffness reduction was taken).

Do not include "leaning columns" in the first-order model. Include "leaning columns" in the Direct Analysis Method model.

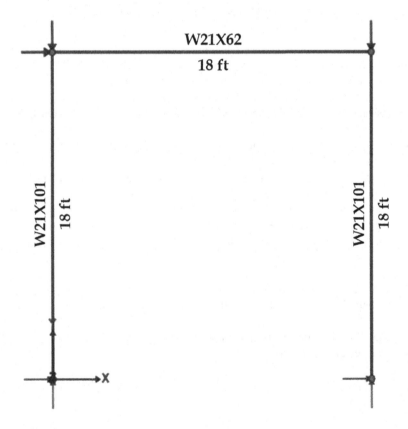

PROBLEM 3.2

Two W40x277 beams (1 shown) spaced 25 ft apart are made from a new Grade 100 steel (F_y = 100 ksi, F_u = 110 ksi) and braced only at the ends and at the loads as indicated by the symbols. The beams are pinned at the left end with a roller at the right end. Loads shown are per beam.

The braces shown are effective for both torsional braces in resisting flexure and for weak axis compression buckling. The braces shown are not effective for strong-axis buckling in compression.

Ignore the beam self-weight effects for hand calculations. Include the self-weight as part of the dead load in the visual analysis model.

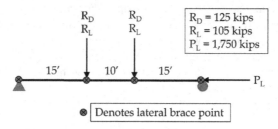

(a) Check the section against Chapter H requirements using a first-order analysis.
(b) Estimate second-order effects using the Direct Analysis Method of Chapter C along with Appendix 8 and recheck the section.
(c) Estimate the strength (factored) displacement due to $D + L$ from a first-order analysis.
(d) Estimate the strength (factored) displacement due to $D + L$ from a second-order analysis.
(e) Model the beam in visual analysis and check the design status and the displacements. Optimize the design of the beam.
(f) Determine the required strength and stiffness of bracing members using Appendix 6. Inverted K Cross frames attached near both flanges are to be used. Design the chords and diagonals.

PROBLEM 3.3

For the plate girder section shown in the figure, determine:

(a) The design flexural resistance in positive bending
(b) The design shear resistance for an end panel

(c) The design shear resistance for an interior panel
(d) The required transverse stiffener size
(e) The required bearing stiffener size if the reaction is equal to the shear
     capacity from part (b)

The girder spans 175 ft and has an unbraced length of 35 ft. C_b for the beam is to
be conservatively taken equal to 1.0. All plates are ASTM A36. Section properties
given are:

- $I_x = 98,399$ in^4
- $I_y = 2,563$ in^4
- $S_x = 2,378$ in^3
- $Z_x = 2,683$ in^3

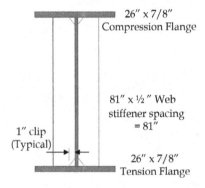

26" x 7/8"
Compression Flange

81" x ½" Web
stiffener spacing
= 81"

1" clip
(Typical)

26" x 7/8"
Tension Flange

PROBLEM 3.4

The frame shown in the figure is to be used as part of the LFRS in a building. The
equivalent lateral force procedure loads are shown in the figure. A992 steel is used.
Axial force in all links is approximately equal to zero. The loads shown in the figure
at each level are from an Equivalent Lateral Force (ELF) seismic analysis by ASCE
7-16 with $R = 8$ incorporated (the given loads have been reduced by $R = 8$ already).
The redundancy factor $\rho = 1.0$. For A992 steel, $R_y = R_t = 1.1$.

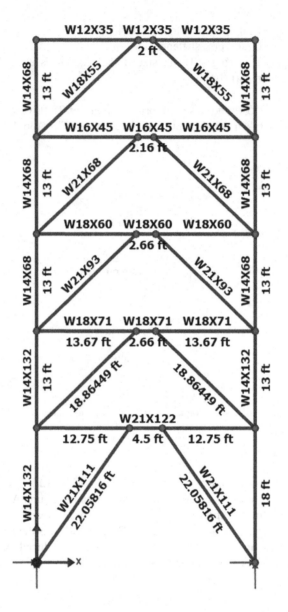

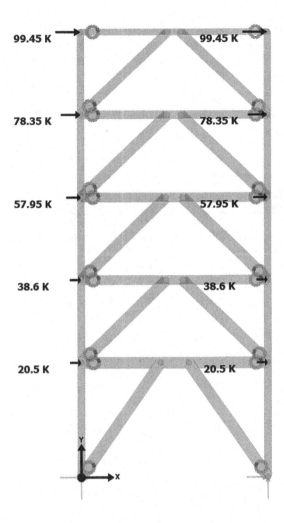

Determine the plastic shear, plastic moment, and link classification (shear, flexural, or intermediate) for each level of the building.

Estimate the link shear, V_u, due to seismic loading and the link resistance, ϕV_n, and determine if each link satisfies strength requirements. All links have highly ductile flanges and highly ductile webs – this does not have to be verified. Summarize the results below.

Given the displacements, δ_{ELF}, in the table below, produced by the *ELF* loading previously shown in Figure P3.4b, determine the amplified plastic drift demand, $\Delta_{P\text{-}EQ}$, and plastic drift capacity, $\Delta_{P\text{-}CAP}$, at each level.

Level	δ_{ELF}, in	Δ_{ELF}, in	$\Delta_{P\text{-}EQ}$, in	$\Delta_{P\text{-}CAP}$, in	OK/NG
Roof	2.006				
5th floor	1.604				
4th floor	1.159				
3rd floor	0.761				
2nd floor	0.411				

Determine the force distribution required for design of the braces and show these loads on the sketch below.

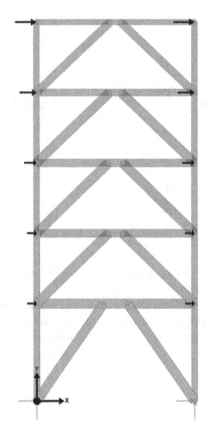

PROBLEM 3.5

For the structure shown, select the lightest members that satisfy the AISC 341 requirements for Special Concentrically Braced Frames.

- Use square HSS sections with A500 Gr. B ($F_y = 46$ ksi) for the braces.
- Use W-shapes with A992 Gr. 50 ($F_y = 50$ ksi) for the beam and columns.
- Assume pin-ended braces and beam.
- Assume that the beam is continuous between the columns.
- Consider the beam unbraced laterally over its entire length.
- The columns are laterally braced at their tops.
- Loads shown are unfactored.
- Lateral loads shown are seismic loads already reduced for the appropriate R value.
- Consider only the $1.2D + 0.5L + 1.0E$ load combination for the brace design.
- Design beams and columns for the capacity-limited E_{cl} loading. This will require some thought since computer software will typically distribute the applied shear equally between the tension and compression braces, while AISC 341-16 requires consideration of unequal distribution between the braces.

$w_D = 1.0$ klf, $w_L = 0.7$ klf
$P_D = 170$ kips, $P_L = 67$ kips, $P_E = 155$ kips

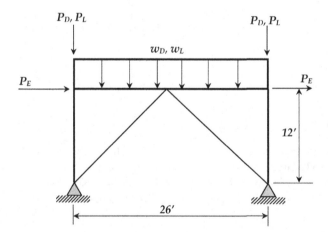

PROBLEM 3.6

For the framing plan shown in the figure, floor loads include:

- Slab and decking – 57 psf
- Framing self-weight – 8 psf
- Mechanical, ceiling, etc. – 10 psf
- Live load – 80 psf
    1. Design a noncomposite floor beam, B1
    2. Design a composite floor beam, B1
    3. Design a noncomposite floor girder, G1
    4. Design a composite floor girder, G1
        – Limit dead load deflection during concrete placement to $L/300$.
        – Limit live load deflection to $L/360$.
        – Camber the beams to 80% of the dead load deflection.
        – For composite designs, assume a 3-inch deck with 3 inches of concrete on top of the deck.

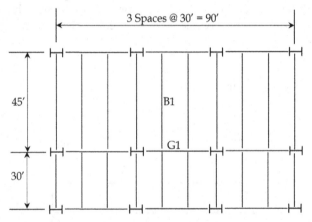

PROBLEM 3.7

The composite plate girder section shown in the figure is in positive bending. All plates are A572 Grade 50 steel. The slab 28-day compressive strength is 6,500 psi. The flanges are 24 inches × 1 inch. The web is 102 inches × 11/16 inch. The haunch (distance from top of beam to bottom of slab) is 2 inches.

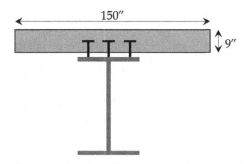

Determine:

(a) The location of the Plastic Neutral Axis (PNA)
(b) The plastic moment of the composite section
(c) The plastic moment of the girder alone

PROBLEM 3.8

For the W18x35 beam "A" shown in the figure, a 3-inch deep metal deck, perpendicular to the beam, with 3 inches of concrete for a 6-inch total depth, is used with $f'_c = 4$ ksi (normal weight concrete). The beam is made from A992 steel and is uniformly loaded. Studs are ¾-inch diameter, with two weak studs per rib. Determine each of the following:

(a) The effective flange width (inches) (I3.1a)
(b) The nominal horizontal shear strength (kips) of one stud, Q_n (Table 3-21)
(c) The required horizontal shear force (kips), ΣQ_n, for fully composite behavior (I3.2d)
(d) The stress block depth (inches), a, for fully composite behavior (page 3-13)
(e) The distance (inches) from the concrete compressive force to the beam top, $Y2$, for fully composite behavior (page 3-13)
(f) The design moment resistance (ft·kips), ϕM_n, for fully composite behavior (Table 3-19)
(g) The total number of studs required for fully composite behavior (I3.2d and I8.2d)
(h) The lower bound moment of inertia (in⁴), I_{LB}, for fully composite behavior (Table 3-20)

(i) The design moment resistance, ϕM_n, for 75% composite behavior (Table 3-19)

(j) The design moment resistance, ϕM_n, for 50% composite behavior (Table 3-19)

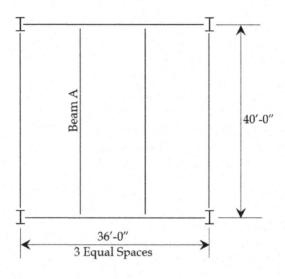

Plan View

PROBLEM 3.9

The W21x44 roof beams on column lines A and D of the four-story AISC building project (Figures 3.7-1 and 3.7-2) are made from A992 steel, as are proposed W14x99 columns of the moment frames on column lines A and D. Dead load shear on the beam end is 12.1 kips. Live load shear on the beam end is 11.5 kips. From an AISC Direct Analysis Method model of the structure, the following loads have been determined:

	LRFD-01	LRFD-02	LRFD-03	LRFD-04
M_u, ft·kips	191.0	95.7	119.6	80.3
V_u, kips	33.0	15.8	19.0	12.2

The building was determined to be Seismic Design Category "B." Note the following excerpt from AISC 341-16, Article A.1:

"User Note: ASCE/SEI 7 (Table 12.2-1, Item H) specifically exempts structural steel systems in seismic design categories B and C from the requirements in these Provisions if they are designed in accordance with the AISC Specification for Structural Steel Buildings and the seismic loads are computed using a seismic response modification coefficient, R, of 3; composite systems are not covered by this exemption. These Provisions do not apply in seismic design category A."

See also Article 14.1.2.2 in ASCE 7-16 for this feature. Since the frame was designed as an ordinary moment frame (OMF) with an R factor for such equal to 3 ½, connections are subject to the requirements of AISC 341-16 and AISC 358-16.

(a) Check the requirements of AISC 341-16 for the beam and column design.
(b) Design a bolted flange plate connection for the beam-to-column as designed: OMF with $R = 3$ ½. Use ¾ inch A325-N bolts.
(c) Discuss the consequences if the frame had been designed using $R = 3$.
(d) Discuss the consequences if the frame had been designed as a special moment frame (SMF).

PROBLEM 4.1

For the steel girder bridge shown, determine the following:

- The location of the center of stiffness
- The anticipated total thermal movement at each substructure
- The required design movements at each substructure
- The thermal force exerted on each pier
- The moment at each column base due to thermal loading

The bridge is located in a moderate climate. The top of each column is pinned at the superstructure. Expansion joints are located at both abutments.

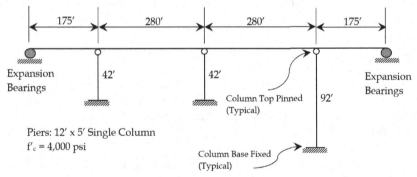

PROBLEM 4.2

A new bridge is to be located at a site with the subsurface profile shown in the figure. Basement rock acceleration values are as follows:

- $PGA = 0.300$

- $S_S = 0.875$
- $S_1 = 0.350$

Determine:

(a) V_{S30}
(b) Site class
(c) Design response spectrum values: A_S, S_{DS}, S_{D1}
(d) Transition periods, T_S and T_O
(e) Seismic zone

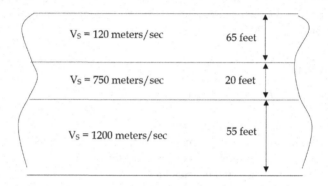

PROBLEM 4.3

A 48-ft wide bridge consists of five girder lines. Welded steel plate girders are to be used. Span lengths are 210 ft, 262 ft 6 inches, and 210 ft. Recommend a beam spacing to be used for the bridge.

PROBLEM 4.4

A two equal span 200-ft long bridge is to be constructed based on the cross section shown in the figure. The substructures are normal to the bridge centerline (no skew). Deck concrete strength is 5 ksi. Welded plate girders are to be used.
    Determine:

(a) The live load distribution factor for moment for an interior girder using the AASHTO equations.
(b) The live load distribution factor for moment for an exterior girder using the lever rule.
(c) The live load distribution factor for moment for an exterior girder using the rigid cross section method.
(d) The live load distribution factor for shear for an interior girder using the AASHTO equations.

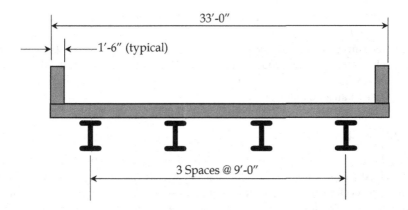

PROBLEM 4.5

All plates are Grade 50W. All bolts are 7/8 inch A325-X. Surface finish is Class B. There are six rows of bolts across the width of the flanges (into the page). Any required filler plates for the field splice are to be the same width as the flange.

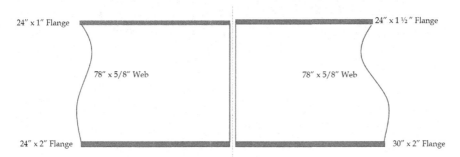

(a) Determine the Strength Limit State design force, P_{fy}, for the top flange splice.
(b) Determine the Strength Limit State design force, P_{fy}, for the bottom flange splice.
(c) Determine the filler plate penalty factor required on Strength Limit State bolt shear resistance for the top flange splice.
(d) Determine the filler plate penalty factor required on Strength Limit State bolt shear resistance for the bottom flange splice.
(e) For the top flange splice plates, a 24-inch × 0.750-inch outer plate with two 10.625-inch × 0.875-inch inner plates is proposed. Do the proposed plates satisfy the AASHTO requirements for equal distribution of the design force to the inner and outer plates? Why or why not?
(f) For the top flange splice plates, a 24-inch × 0.750-inch outer plate with two 10.625-inch × 0.875-inch inner plates is proposed. Do the proposed plates

satisfy the minimum area requirement based on the flange size? Why or why not?

(g) Determine the *total* number of bolts required for the top flange splice based solely on bolt shear. You do not have to check bearing, tear-out, or slip for this exam problem even though you would be required to check each of these in an actual design.

PROBLEM 4.6

Given the composite beam in the figure, suppose a transverse stiffener is to be added and welded to the web and to both flanges. Assess the adequacy of a transverse stiffener-to-bottom flange weld for fatigue. Positive moment causes tension in the bottom flange. The location under question is near mid-span of span one in a two-span continuous beam bridge. The $ADTT_{SL}$ is 1,800 trucks per day.

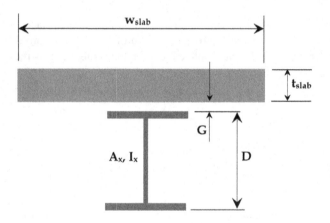

The girder moments are:

$M_{LL} = 330$ ft·kips, maximum positive
$M_{LL} = -22$ ft·kips, maximum negative

These are live load moments and the values shown do not include impact or load factors. The beam is a W36x150 made from A709 Grade 50W steel.

$w_{slab} = 9$ ft $f'_c = 4.5$ ksi $t_{slab} = 8.5$ inches
$G = 2$ inches $E_c = 3,860$ ksi

Beam properties:

$A = 44.3$ in² $I_x = 9,040$ in⁴
$D = 35.9$ inches $t_f = 0.940$ inches

The modular ratio, n, is to be taken equal to 8.0. Assume that the concrete area between the top of the beam and the bottom of the deck is ineffective, but that the gap, G, still exists.

The interstate bridge has four lanes, all available to trucks, and is in a rural area. Determine:

(a) The transformed slab width for short-term, elastic, composite property calculations
(b) The short-term, elastic, composite neutral axis location from the bottom of the beam
(c) The short-term composite moment of inertia
(d) The short-term, elastic, composite section modulus for the bottom of the beam
(e) The factored stress range, $\gamma \Delta f$, in the bottom of the beam for the Fatigue 1 Limit State
(f) The fatigue category for the detail in question
(g) For Fatigue II Limit State evaluation, the number of cycles, N
(h) For Fatigue II Limit State evaluation, the limiting stress range, $(\Delta F)_n$
(i) For Fatigue I Limit State evaluation, the limiting stress range, $(\Delta F)_n$

PROBLEM 4.7

The composite steel welded plate girder shown in the figure has the properties indicated in the accompanying table. "y-elastic" and "y-plastic" are the distances from the bottom of the girder to the neutral axis for the elastic and plastic states, respectively. The distance from the top of the girder to the bottom of the deck is 2 inches.

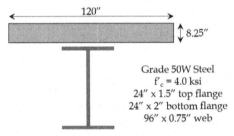

120"

8.25"

Grade 50W Steel
$f'_c = 4.0$ ksi
24" x 1.5" top flange
24" x 2" bottom flange
96" x 0.75" web

	Girder	Girder w/ Rebar	Composite (n)	Composite ($3n$)
A, in^2	156.00	165.90	279.75	197.25
I, in^4	253,838	288,423	510,915	375,273
S_{top}, in^3	4,760	5,804	19,379	9,248
S_{bott}, in^3	5,498	5,790	6,986	6,369
M_p, ft·kips	24,113	27,140	36,086	36,086
y-elastic, in	46.17	49.81	73.14	58.92
y-plastic, in	42.00	49.92	86.88	86.88

Required: For long-term loads, determine the following:

(a) The elastic depth of web in compression, D_c, in positive bending
(b) The plastic depth of web in compression, D_{cp}, in positive bending
(c) The elastic depth of web in compression, D_c, in negative bending
(d) The plastic depth of web in compression, D_{cp}, in negative bending
(e) The nominal moment resistance, M_n, of the girder in positive bending at the Strength Limit State

PROBLEM 4.8

The bridge cross section shown consists of steel plate girders with 24-inch wide flanges and a 72-inch deep web. Determine the live load distribution factor for an exterior girder using the rigid cross section method (Equation 4.7-8). Consider all lane cases up to the maximum.

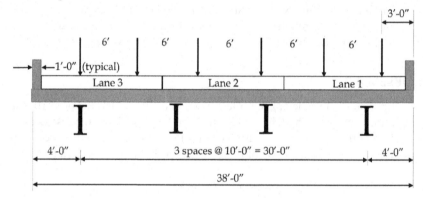

PROBLEM 5.1

For the Caruthersville Bridge, Interstate 155, over the Mississippi River in Tennessee, determine the coordinates of the bridge and a value for inferred shear wave velocity at the site. Determine uniform hazard, geometric mean-based 2,500-year-*MRI PGA*, S_S, and S_1 values. Determine modal M, R combinations for *PGA*, S_S, and S_1.

PROBLEM 5.2

Using the ground motions summarized in the table below:

(a) Generate 5% damping *PSA* spectra for each of the records and compute geometric mean *PSA* of the two horizontal components for each record pair. Use a period range of 0.00–6.00 sec.
(b) Compute log-based standard deviation of *PSA*. Use a period range of 0.00–6.00 sec.

(c) Plot the average *PSA* spectrum for the 11 record pairs. On the same plot, include the target spectrum defined by SDS = 1.018, SD1 = 0.786, and TL = 8 seconds.

(d) Plot the log-based standard deviation of *PSA*. Use a period range of 0.00– 6.00 sec.

(e) Estimate mean significant duration for each of the records. Compute the average $D_{5\text{-}95}$ for the suite. Compute the log-based standard deviation of $D_{5\text{-}95}$.

(f) Estimate median *PGA*, *PGV*, and *PGD* for the records and compute averages for each.

(g) Compute log-based standard deviation for each.

(h) Estimate median + one Sigma $D_{5\text{-}95}$, *PGA*, *PGV*, and *PGD*.

PEER RSN	Scale Factor	Details
6890-N10E	2.4128	Darfield New Zealand, Christchurch Cashmere High School, N10E
6890-S80E	2.4128	Darfield New Zealand, Christchurch Cashmere High School, S80E
6952-S33W	4.6051	Darfield New Zealand, Papanui High School , S33W
6952-S57E	4.6051	Darfield New Zealand, Papanui High School , S57E
5823-000	2.8012	El Mayor-Cucapah, Chihuahua, 0
5823-090	2.8012	El Mayor-Cucapah, Chihuahua, 90
5832-000	2.5031	El Mayor-Cucapah, TAMAULIPAS, 0
5832-090	2.5031	El Mayor-Cucapah, TAMAULIPAS, 90
5859-360	2.4507	El Mayor-Cucapah, Westmorland Fire Sta, 360
5859-090	2.4507	El Mayor-Cucapah, Westmorland Fire Sta, 90
0573-090	2.2932	TAIWAN SMART1 (45) , SMART1 I01, EW
0573-180	2.2932	TAIWAN SMART1 (45) , SMART1 I01, NS
1147-000	3.0633	KOCAELI , AMBARLI, N (KOERI)
1147-090	3.0633	KOCAELI , AMBARLI, E (KOERI)
1155-000	2.2176	KOCAELI , BURSA TOFAS, N (KOERI)
1155-090	2.2176	KOCAELI , BURSA TOFAS, E (KOERI)
1203-000	3.1933	CHI-CHI , CHY036, N
1203-090	3.1933	CHI-CHI , CHY036, E
1204-000	2.6169	CHI-CHI , CHY039, N
1204-090	2.6169	CHI-CHI , CHY039, E
3758-135	1.5227	Landers, Thousand Palms Post Office, 135
3758-045	1.5227	Landers, Thousand Palms Post Office, 45

PROBLEM 5.3

Develop the design response spectrum (DRS) for a Seismic Design Category (SDC) 5 project in Hanford, Washington, at the former nuclear operations site in accordance with ASCE 43-05.

(a) Locate the coordinates of the Hanford site using Google Maps or some other mapping tool.

(b) Obtain the mapped bedrock acceleration values, S_S and S_1, for all necessary annual probability of exceedance values, using the appropriate USGS online tools.

(c) Determine an inferred shear wave velocity, V_{S30}, using OpenSHA. Use this to determine a site class in accordance with ASCE 7-16.

(d) Propagate the bedrock accelerations to the surface for each hazard level using ASCE 7-16 site factors, F_a and F_v.

(e) Determine design factors, DF, at each period (PGA, 0.2 sec, and 1.0 sec).

(f) Compute the DRS from the UHRS parameters and design factors.

(g) Plot the bedrock and surface spectra on the same log-based chart.

(h) Summarize all results and plots in a single Excel file.

PROBLEM 5.4

An emergency management operations building project at Dyersburg State Community College in Dyersburg, TN, requires nonlinear response history analysis for final design. Develop the MCER target response spectrum and select a suite of ground motion record pairs for the analysis.

(a) Locate the coordinates of the project using Google Maps or some other mapping tool.

(b) Use OpenSHA to obtain an inferred shear wave velocity for the site.

(c) Determine the appropriate site class.

(d) Use the USGS Unified Hazard Tool to determine GeoMean-based S_{S-UHS} and S_{1-UHS} at bedrock.

(e) Multiply by the appropriate factors to convert *GeoMean*-based values for S_{S-UHS} and S_{1-UHS} at bedrock to *RotD100*-based values.

(f) Use the ATC Hazard by Location online tool to determine *RotD100*-based S_S and S_1 values (to verify values found in the next step), as well as risk coefficients, C_{RS} and C_{R1}.

(g) Apply C_{RS} and C_{R1} to *RotD100*-based S_{S-UHS} and S_{1-UHS} to obtain *RotD100*-based risk-targeted ground motion parameters, S_S and S_1. Verify the values reported by the ATC application.

(h) Apply the appropriate site factors, F_a and F_v, from ASCE 7-16 to propagate the bedrock S_S and S_1 to the surface.

(i) Determine the long period transition period, T_L, from the ATC Hazard by Location App.

(j) Develop the *RotD100*-based, risk-targeted, surface MCE$_R$ target *PSA* spectrum up to a period of 5.0 sec.

(k) Establish an appropriate period range for ground motion selection and modification criteria. The natural periods in the two horizontal directions are 0.80 and 1.35 sec. The period at which at least 90% of the total mass has been accounted for is 0.30 sec.

(l) Disaggregate the seismic hazard using USGS online tools to determine the modal and mean M, R, combinations.

(m) Use the PEER Ground Motion Database or *SigmaSpectra* to select and scale a suite of 11 ground motion records for the project.

(n) Summarize all findings, results, and plots in a single Excel file.

Appendix A
Hand Calculations for Example 2.3-1

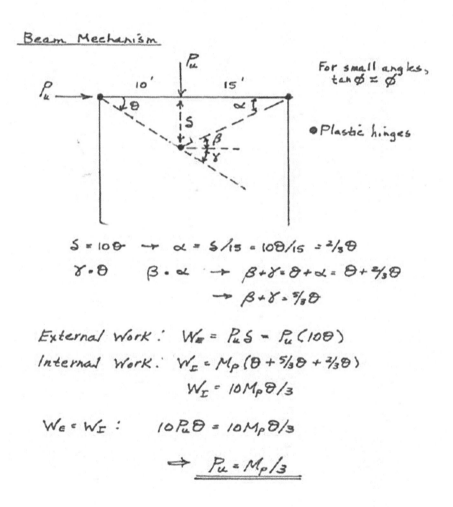

Beam Mechanism

For small angles,
$\tan \phi \approx \phi$

● Plastic hinges

$S = 10\theta \longrightarrow \alpha = S/15 = 10\theta/15 = 2/3\theta$

$\gamma = \theta \qquad \beta = \alpha \longrightarrow \beta + \gamma = \theta + \alpha = \theta + 2/3\theta$

$\longrightarrow \beta + \gamma = 5/3\theta$

External Work: $W_E = P_u S = P_u (10\theta)$

Internal Work: $W_I = M_p (\theta + 5/3\theta + 2/3\theta)$

$W_I = 10 M_p \theta / 3$

$W_E = W_I :\qquad 10 P_u \theta = 10 M_p \theta / 3$

$\Longrightarrow \underline{\underline{P_u = M_p / 3}}$

Sway Mechanism

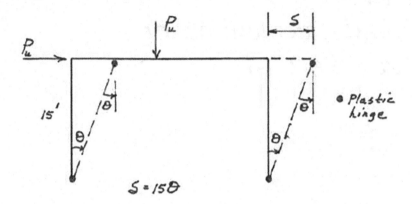

$$S = 15\theta$$

External Work: $W_E = P_u S = 15 P_u \theta$

Internal Work: $W_I = M_p (\theta + \theta + \theta + \theta)$

$W_E = W_I :$ $15 P_u \theta = 4 M_p \theta$

$$\Rightarrow \quad \underline{\underline{P_u = 4/15 \, M_p}}$$

Combined Mechanism

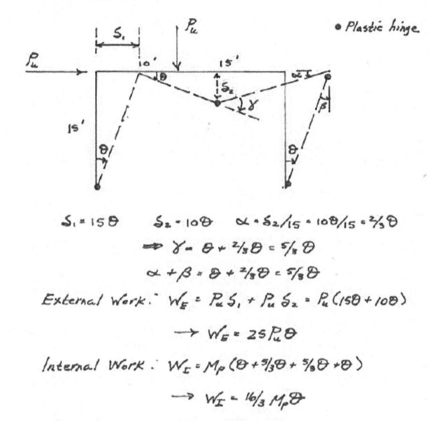

$$S_1 = 15\theta \qquad S_2 = 10\theta \qquad \alpha = S_2/15 = 10\theta/15 = \tfrac{2}{3}\theta$$

$$\implies \gamma = \theta + \tfrac{2}{3}\theta = \tfrac{5}{3}\theta$$

$$\alpha + \beta = \theta + \tfrac{2}{3}\theta = \tfrac{5}{3}\theta$$

External Work: $W_E = P_u S_1 + P_u S_2 = P_u(15\theta + 10\theta)$

$$\longrightarrow W_E = 25 P_u \theta$$

Internal Work: $W_I = M_p(\theta + \tfrac{2}{3}\theta + \tfrac{5}{3}\theta + \theta)$

$$\longrightarrow W_I = \tfrac{16}{3} M_p \theta$$

$W_E = W_I :$ $\qquad 25 P_u \theta = \tfrac{16}{3} M_p \theta$

$$\implies \underline{\underline{P_u = 16/75\, M_p}}$$

Use statics to work your way around the frame:

Column 2: $P = 208^k$ $V = 208^k$ $L_b = 15'$

$M_2 = M_p = 1,554 \; ft \cdot k$
$M_1 = -M_p = 1,554 \; ft \cdot k$

Beam 2: $P = 208^k$ $V = 208^k$ $L_b = 15'$

$M_2 = M_p = 1,554 \; ft \cdot k$
$M_1 = -M_p = -1,554 \; ft \cdot k$

Beam 1: $P = 208^k$ $V = 124^k$ $L_b = 10'$

$M_2 = M_p = 1,554 \; ft \cdot k$
$M_1 = 314 \; ft \cdot k$

Column 1: $P = 124^k$ $V = 124^k$ $L_b = 15'$

$M_2 = M_p = 1,554 \; ft \cdot k$
$M_1 = -314 \; ft \cdot k$

<u>AISC 360-16, Appendix 1 checks:</u>

1) $F_y = 50 \; ksi < 65 \; ksi$, OK

2) W21×147 $A = 43.2 \; in^2$

$P_y = 50 \times 43.2 = 2,160 \; KIPS$

$\dfrac{P_u}{\phi_c P_y} = \dfrac{208}{.9 \times 2,160} = 0.107 < 0.125$
                                         (worst case P_u)

$(\lambda_{pd})_{web} = 3.76 \sqrt{\dfrac{29,000}{50}} \, (1 - 2.75 \times .107) = 60.6$

$h/t_w = 26.1 < 60.6$, Web OK.

$(\lambda_{pd})_{flange} = 0.38 \sqrt{\dfrac{29,000}{50}} = 9.15$

$b/t = b_f/2t_f = 5.44 < 9.15$,
                                         Flange OK

3) Beam 1 : $L_b = 10 ft = 120 in.$

$\qquad$ $W21 \times 147$, $r_y = 2.95''$

$\qquad$ $M_2 = 17,274 \ in-k$

$\qquad$ $M_1 = 2,223 \ in\cdot k$

$\qquad$ M_2 is always positive. M_1 is positive when it causes compression in the same flange as does M_2.

$\qquad$ With linear moment diagrams,
$\qquad$ $M_{mid} = \frac{1}{2}(M_1 + M_2)$

$\qquad$ $\longrightarrow$ Case (2) $\longrightarrow$ $M_1' = M_1$

$\qquad$ $L_{pd} = \left[0.12 - 0.076\left(\frac{314}{1,554}\right)\right]\frac{29,000}{50}(2.95)$

$\qquad$ $L_{pd} = 179''$

$\qquad$ $L_b = 120'' < L_{pd} = 179'' \longrightarrow ok$

$\quad$ Beam 2 : $L_b = 15' = 180''$

$\qquad$ $M_1' = -M_p$ $M_2 = M_p$

$\qquad$ $L_{pd} = \left[0.12 - 0.076\left(-1.00\right)\right]\frac{29,000}{50}(2.95)$

$\qquad$ $L_{pd} = 335''$

$\qquad$ $L_b = 180'' < L_{pd} = 335'' \longrightarrow ok$

$\quad$ Column 1 : $L_b = 15' = 180''$

$\qquad$ $M_1' = -314$ $M_2 = 1,554$

$\qquad$ $L_{pd} = \left[0.12 - 0.076\left(\frac{-314}{1,554}\right)\right]\frac{29,000}{50}(2.95)$

$\qquad$ $L_{pd} = 232''$

$\qquad$ $L_b = 180'' < L_{pd} = 232'' \longrightarrow ok$

Column 2 : $L_b = 15' = 180''$

$\qquad M_1 = -M_p \qquad M_2 = M_p$.

$\qquad L_{pd} = \left[0.12 - 0.076 \left(\dfrac{M_p}{M_p} \right) \right] \dfrac{29,000}{50} (2.95)$

$\qquad L_{pd} = 335''$

$\qquad L_b = 180'' < 335'' \longrightarrow OK$

4) $P_u \leq 0.75 F_y A_g$

$\qquad 0.75 \times 50 \times 43.2 = 1,620 \, kips$

$\qquad All \ axial \ loads \ are \ < 1,620 \, kips$

$\qquad \longrightarrow OK$

The frame meets all 4 criteria. So plastic analysis is applicable.

Appendix B
Hand Calculations for Example 3.3-1

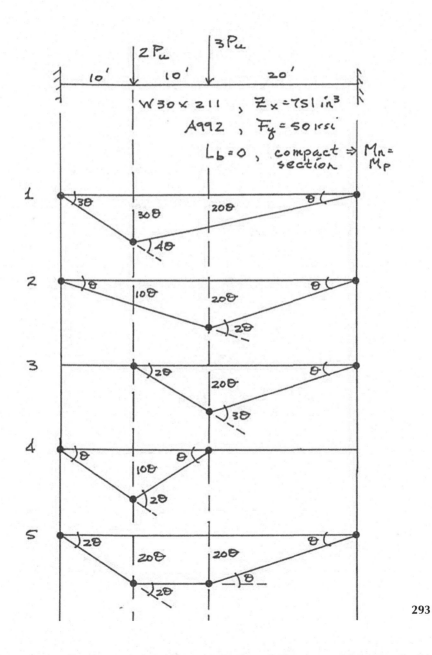

1. $W_I = M_p(3\theta + 4\theta + \theta) = 8\,M_p\theta$

   $W_E = 2P_n\,(30\theta) + 3P_n\,(20\theta) = 120\,P_n\,\theta$

   $8M_p\theta = 120\,P_n\theta \implies \underline{P_n = \dfrac{8}{120}\,M_p = \dfrac{M_p}{15}}$

2. $W_I = M_p(\theta + 2\theta + \theta) = 4M_p\theta$

   $W_E = 2P_n\,(10\theta) + 3P_n\,(20\theta) = 80\,P_n\theta$

   $4M_p\theta = 80\,P_n\theta \implies \underline{P_n = \dfrac{4}{80}\,M_p = \dfrac{M_p}{20}}\;\Leftarrow$

3. $W_I = M_p(2\theta + 3\theta + \theta) = 6M_p\theta$

   $W_E = 2P_n\,(0) + 3P_n\,(20\theta) = 60\,P_n\theta$

   $6M_p\theta = 60\,P_n\theta \implies \underline{P_n = \dfrac{6}{60}\,M_p = \dfrac{M_p}{10}}$

4. $W_I = M_p(\theta + 2\theta + \theta) = 4M_p\theta$

   $W_E = 2P_n\,(10\theta) + 3P_n\,(0) = 20\,P_n\theta$

   $4M_p\theta = 20\,P_n\theta \implies \underline{P_n = \dfrac{4}{20}\,M_p = \dfrac{M_p}{5}}$

5. $W_I = M_p(2\theta + 2\theta + \theta + \theta) = 6M_p\theta$

   $W_E = 2P_n\,(20\theta) + 3P_n\,(20\theta) = 100\,P_n\theta$

   $6M_p\theta = 100\,P_n\theta \implies \underline{P_n = \dfrac{6}{100}\,M_p = \dfrac{M_p}{16\,{}^2\!/_3}}$

$M_p = F_y Z_x = 50 \times 751 = 37{,}550 \text{ in-k} = 3{,}129 \text{ ft·k}$

$P_n = 3{,}129/20 = 156.5 \text{ kips}$

$\underline{\underline{\phi P_n = .9 \times 156.5 = 140.8 \text{ kips}}}$

Appendix C
Hand Calculations for Example 3.5-2

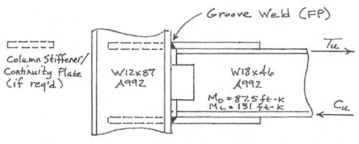

Groove Weld (FP)

Column Stiffener/
Continuity Plate
(if req'd)

W12×87
A992

W18×46
A992

$M_D = 87.5$ ft-k
$M_L = 131$ ft-k

T_u

C_u

A992 , $F_y = 50$ ksi , $F_u = 65$ ksi

W12×87 Column , $k = 1.41''$ $d = 12.5''$
$t_f = 0.81''$ $t_w = 0.515''$
$b_f = 12.1''$

W18×46 Beam , $b_f = 6.06''$ $d = 18.1''$ $t_f = 0.605''$

$M_u = 1.2 \times 87.5 + 1.6 \times 131 = 314.6$ ft-kips

Distance between beam flange centroids

$= 18.1 - 0.605 = 17.5''$

$C_u = T_u = \dfrac{314.6 \text{ ft-k} \times 12''/\text{ft}}{17.5} = 216$ kips

Column Flange Local Bending (J10.1)

$\phi R_n = 0.90 (50\text{ksi})(0.81'')^2 (6.25)$

$\Rightarrow \phi R_n = 184$ kips $< T_u = 216$ kips

∴ Top column stiffener is req'd
(or choose a column with
thicker flanges).

Column Web Local Yielding (J10.2)

Check assuming that this connection
is at the roof, near the end of
the column.

$$\phi R_n = 1.00 \,(50\,ksi)(0.515")(2.5\times1.41+1.41)$$

$$\Rightarrow \phi R_n = 127 \text{ kips} < T_u = 216 \text{ kips, No Good}$$

$$\Rightarrow \text{Stiffener pairs are req'd}$$
$$\text{at both flanges (or doubler plate)}$$

Column Web Local Crippling (J10.3)

$$Q_f = 1.0 \qquad d = 12.5" \qquad t_w = 0.515" \qquad t_f = 0.81"$$

$$l_b = 1.41"$$

$$\phi R_n = 0.75\,(0.80)(0.515)^2\left[1+3\left(\frac{1.41}{12.5}\right)\left(\frac{0.515}{0.81}\right)^{1.5}\right]\sqrt{\frac{29,000\times50\times0.81}{0.515}}\,(1.0)$$

$$\Rightarrow \phi R_n = 281 \text{ kips} > C_u = T_u = 216 \text{ kips}$$

Since this req't is satisfied, the ¾ d_w stiffener extension at the end of J10.3 does not apply.

Web Sidesway Buckling (J10.4)   not applicable

Web Compression Buckling (J10.5)   not applicable

Web Panel Zone Shear (J10.6)   not enough info given (See Commentary)

Stiffener Req'ts (J10.8)

Stiffener to: either bear on or be welded to the loaded flange

be welded to the web

$$b_s + \tfrac{1}{2}(0.515) \geq \tfrac{1}{3}(b_f)_{beam} = \tfrac{1}{3}(6.06")$$

$$\Rightarrow b_s \geq 2.02" - \tfrac{1}{2}(0.515") = 1.76"$$
$$\Rightarrow \text{take } b_s = 2"$$

$$t_s \geq \tfrac{1}{2}\,(0.605)^* = 0.303" \,,\quad t_s \geq \tfrac{2}{16} = 0.125"$$

Use 2" × 5/16" stiffener pairs at both flanges, 7" long stiffeners

* (We have assumed the connection plates will be the same thickness as the beam flanges, which won't always be true.)

Appendix D
Hand Calculations for Example 3.5-3

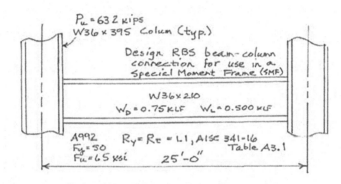

$P_u = 632$ kips
W36×395 Column (typ.)

Design RBS beam-column
connection for use in a
Special Moment Frame (SMF)

W36×210
$W_D = 0.75$ KLF $W_L = 0.500$ KLF

A992 $R_y = R_t = 1.1$, AISC 341-16
$F_y = 50$ Table A3.1
$F_u = 65$ ksi   25'-0"

Column: $d = 38.4"$ $b_f = 16.8"$ $A_g = 116$ in²
$t_w = 1.22"$ $t_f = 2.20"$ $Z_x = 1,710$ in³
$k = 3.15"$

Beam: $d = 36.7"$ $b_f = 12.2"$ $Z_x = 833$ in³
$t_w = 0.83"$ $t_f = 1.36"$ $r_y = 2.58$ in
    $I_y = 411$ in⁴

<u>AISC 358-16. RBS Moment Connection Design</u>

5.3.1. Beam Limits   W36 max, OK
wt. 302 plf max, OK
$t_{fb} = 1.36" < 1.75"$, OK

clear span = 25' - 38.4/12 = 21.8'

clear span-to-depth = 21.8×12/36.7
    = 7.13 > 7, OK

b/t & h/t_w limits from AISC 341-16 (not 360-16)

AISC 341-16, E3.5a, beam & column must be
    "highly ductile"

AISC 341-16, Table D1.1:

$$\lambda_{hd} = 0.32 \sqrt{\frac{29,000}{1.1 \times 50}} = 7.35, \text{ flange}$$

$$\lambda_{hd} = 2.57 \sqrt{\frac{29,000}{1.1 \times 50}} = 59.0, \text{ web (beam)}$$

$C_a = \dfrac{632}{.9 \times 1.1 \times 50 \times 116} = 0.11 \to \lambda_{hd} = 59.0(1-1.04 \times 0.11) = 52.2$, web (col.)

Beam $b/t = 4.48 < 7.35$, OK (flange)

$h/t_w = 39.1 < 59.0$, OK (web)

Column $b/t = 3.83 < 7.35$, OK (flange)

$h/t_w = 26.3 < 52.2$, OK (web)

Beam lateral bracing : AISC 341-16 , D1.2b

$$L_b \leq \frac{0.095(2.58)(29,000)}{1.1 \times 50} = 129" = 10.8'$$

5.3.2. Column Limits   W36 max, OK
See above for λ_{hd} req'ts

5.8 Design Procedure

RBS dimensions : $0.5(12.2) \leq a \leq 0.75(12.2)$
$6.1" \leq a \leq 9.15"$

$\underline{\text{Select } a = 9"}$

$0.65(36.7) \leq b \leq 0.85(36.7)$
$23.85" \leq b \leq 31.2"$

$\underline{\text{Select } b = 24"}$

$0.1(12.2) \leq c \leq 0.25(12.2)$
$1.22" \leq c \leq 3.05"$

$\underline{\text{Select } c = 2"}$

RBS - Z : $Z_{RBS} = 833 - 2(2")(1.36")(36.7 - 1.36)$

$\Rightarrow \underline{Z_{RBS} = 641 \text{ in}^3}$

RBS - M_{pr} : $C_{pr} = \frac{50+65}{2(50)} = 1.15$ (2.4.3)

$M_{pr} = 1.15 \times 1.1 \times 50 \times 641$

$\Rightarrow \underline{M_{pr} = 40,527 \text{ in-k}}$
$\underline{= 3,377 \text{ ft-k}}$

RBS - Shear Force :

$$W_u = 1.2 \times 0.75 + 0.5 \times 0.500 = 1.15 \text{ KLF}$$

distance from RBS-to-RBS
$$= 25' - 38.4/12 - 2(a + b/2)$$
$$= 25' - 3.20' - 2(9 + 12)/12$$
$$= 18.3 \text{ ft}$$

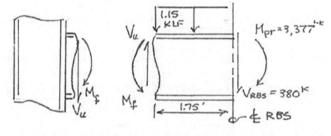

$M_{pr} = 3,377^{\,\prime-k}$ (1.15 KLF   ) $M_{pr} = 3,377^{\,\prime-k}$

18.3'

$V_1 = V_{RBS}$ V_2

$$V_{RBS} = \frac{2(3,377) + 1.15(18.3 \times 18.3/2)}{18.3}$$

$$= 369^k + 11^k = \underline{380 \text{ kips}}$$

Column face Beam Moment, M_f

$$S_h = a + b/2 = 9'' + 12'' = 21'' = 1.75'$$

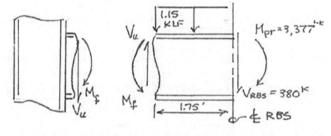

$$M_f = 3,377 + 380(1.75) + 1.15(1.75)(1.75/2)$$

$$\Rightarrow \underline{M_f = 4,044 \text{ ft} \cdot k}$$

$$V_u = 380 + 1.15(1.75)$$

$$\Rightarrow \underline{V_u = 382 \text{ kips}}$$

Beam M_{pe} : $M_{pe} = 1.1 \times 50 \times 833 = 45,815$ in-k
$\qquad\qquad\qquad\qquad\qquad = 3,818$ ft·k

$\phi_d = 1.00$, 2.4.1

$\phi_d M_{pe} = 1.00 \times 3,818 = 3,818$ ft·k

$M_f = 4,044$ ft·k $\nleq \phi M_{pe} \Rightarrow$ No Good

Hinge forms at beam end, not at RBS.

Adjust dimensions :
$\quad\begin{array}{l} a = 7'' \\ b = 24'' \\ c = 3'' \end{array}$

$\Rightarrow Z_{RBS} = 545$ in³

$\quad M_{pr} = 2,871$ ft·k

$\quad V_{RBS} = 308 + 11 = 319$ kips

$\quad R_{BS} - to \cdot RBS$ span $= 18.63$ ft

$\quad S_h = 1.58'$

$\quad M_f = 3,376$ ft·k $< \phi_d M_{pe} = 3,818$ ft·k

$\qquad\qquad\qquad\qquad\qquad \Rightarrow$ OK

Hinge forms at RBS, not at beam end.

$V_u = 321$ kips

Beam-to-Column Connection
Design Forces :
$\qquad\qquad M_f = 3,376$ ft·k
$\qquad\qquad V_u = 321$ kips

Beam Shear Resistance :

$\phi V_n = 914$ kips
$> V_u = 321$ kips
ok

AISC Manual
Page 3-21

5.4   Column-Beam Moment Ratio

$$M_{uv} = V_{RBS}(a + b/2 + d_c/2)$$

$$= 319(7 + 12 + 38.4/2)$$
$$= 12,186 \text{ in-k}$$
$$= 1,015 \text{ ft-k}$$

AISC 341-16, E3.4a, $\dfrac{\Sigma M_{pc}^*}{\Sigma M_{pb}^*} > 1.0$

$$M_{pc}^* = 1,710(50 - 632/16)$$
$$= 76,183 \text{ in-k}$$
$$= 6,349 \text{ ft-k}$$

$$M_{pb}^* = 2,871 + 1,015 = 3,886$$

With RBS on one side of column only:

$$\frac{\Sigma M_{pc}^*}{\Sigma M_{pb}^*} = \frac{2 \times 6,349}{3,886} = 3.27 > 1.0$$
$$OK$$

With RBS on both sides:

$$\frac{\Sigma M_{pc}^*}{\Sigma M_{pb}^*} = \frac{2 \times 6,349}{2 \times 3,886} = 1.63 > 1.0$$
$$OK$$

With RBS on both sides, top of bldg:

$$\frac{\Sigma M_{pc}^*}{\Sigma M_{pb}^*} = \frac{6,349}{2 \times 3,886} = 0.82 < 1.0$$
$$No \; Good$$

Continuity Plate Req'ts , AISC 341-16 , E3.6e

$$R_u = 2 V_u \text{ , with RBS on both sides of column}$$

$$R_u = 2 \times 321^K = 642 \text{ kips}$$

ϕR_n , Panel Zone , AISC 360-16 , J10.6

$$\frac{P_u}{P_y} = \frac{632}{50 \times 116} = 0.11$$

$$R_n = 0.6 F_y d_c t_w$$

$$= 0.6 \times 50 \times 38.4 \times 1.22$$

$$= 1,405 \text{ kips}$$

$$\phi - 1.00 , \text{ AISC 341-16, E3.6e}$$

$$\phi R_n = 1.00 \times 1,405 = 1,405^K > V_u = 642^K$$

OK , doubler plates not req'd

AISC 341-16 , E3.6e

$$t \geq (d_z + w_z)/90$$

$$d_z = 36.7 - 2(1.36) = 33.98''$$

$$W_z = \text{width of column panel zone}$$
$$= 38.4 - 2(2.20) = 34.00''$$

$$t \geq (33.98 + 34.00)/90 = 0.755''$$

$$t_w = 1.22'' > 0.755'' , \text{ ok}$$

AISC 360-16 , J10.1 , Flange Local Bending

$$R_n = 6.25 (50)(2.20)^2 = 1,512^K$$
$$\phi R_n = .9 \times 1,512 = 1,361 \text{ kips}$$

$$R_u = \frac{M_f}{(d - t_f)_{beam}} = \frac{3,376 \times 12}{36.7 - 1.36} = 1,146^K$$

$$1,146^K < 1,361^K \rightarrow \text{ ok}$$

Appendix E
Hand Calculations for Example 4.2-1

Check stress with Bearing Fully extended:

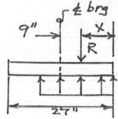

$X = 27/2 - 9 = 4.50''$

$f_s = \dfrac{498^K}{(2 \times 4.50)(28)}$

$\rightarrow \underline{f_s = 1.98 \text{ ksi} < 2.0 \text{ ksi OK}}$

Bearing on Concrete   (AASHTO 5.6.5)

$\phi P_n = 0.70 [0.85 f'_c A_1 m]$

conservatively take $m = 1$

$A_1 = brg$ area $= BW \times BL$

$BW = 29''$ $BL = 36''$

check with slider fully extended:

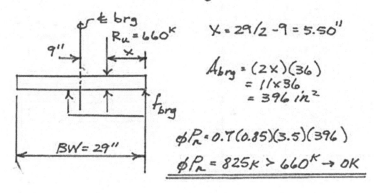

$X = 29/2 - 9 = 5.50''$

$A_{brg} = (2x)(36)$
$= 11 \times 36$
$= 396 \text{ in}^2$

$\phi P_n = 0.7(0.85)(3.5)(396)$
$\underline{\phi P_n = 825^k > 660^k \rightarrow OK}$

Shear in Anchor Rods   (AASHTO 6.13.2.12)

$1\frac{1}{2}"\ \phi$ F1554 Grade 105 anchor rod

$\phi R_n = 0.75 \, [\, 0.50 A_b \, F_{ub} \, N_s \,]$

$\qquad A_b = \pi (1.5)^2 / 4 = 1.767 \ in^2 / anchor$

$\qquad F_{ub} = 125 \, ksi \quad for \ Grade \ 105 \ anchors$

$\phi R_n = 0.75 \, [\, 0.50 \times 1.767 \times 125 \times 1 \,]$

$\qquad \Rightarrow \phi R_n = 82.8 \, k / anchor$

$\qquad\qquad\quad \underline{\times \ 4 \ anchors}$

$\qquad\qquad \underline{\underline{\phi R_n = 331 \, kips}}$

$R_u = \mu W = 0.4 \times 660 \Rightarrow R_u = 264^k < \phi R_n = 331^k$

$\qquad\qquad\qquad\qquad \underline{\underline{\longrightarrow anchors \ ok}}$

(In case base plate moves
    before slider does)

Appendix F

Hand Calculations for Example 3.9-1

AISC 360-16, J8

$$\phi P_p = 0.65\,(0.85\,f'_c)\,A_1$$

Determine whether anchor tension is required for equilibrium:

$$e = M_u/P_u = 610(12)/610$$
$$\Rightarrow \underline{e = 12.00''}$$

For $T = 0$, centroid of $\mathcal{g}$ coincides with location of P_u

$$\Rightarrow Y_{T=0} = 2 \times 3.00''$$
$$= 6.00''$$

$$(\phi P_p)_{T=0} = 0.65(0.85)(6.5)(6 \times 15)$$
$$= 323^K < P_u = 610^K$$

$\Rightarrow$ Anchor tension is req'd to maintain equilibrium.

$$\Sigma F_y = 0: \quad P_u + T = g_{max}(Y)(15)$$

$$g_{max} = 0.65(0.85)(6.5) = 3.59\ ksi$$

$$P_u + T = 3.59\,Y(15)$$

$$610 + T = 53.87\,Y$$

$$\Sigma M_A = 0: \quad 610(12 + 15 - 3) = 53.87\,Y(2T - Y/2)$$

$$14,640 = 1,454.456\,Y - 26.934\,Y^2$$

$$\Rightarrow \underline{Y = 13.38''} \quad \Rightarrow \underline{T = 111\ kips}$$

AISC 360-16, J9

    Table J3.2 $\Rightarrow$ $F_{nt} = 0.75\, F_u$
                      $F_{nv} = 0.563\, \overline{F_u}$

    J3.6, $\phi R_n = 0.75\,(F_{nt})\, A_b$

    F1554 Anchor Rod, Page 2-52
           Grade 36, $\overline{F_u} = 58\,ksi$
           Grade 55, $F_u = 75\,ksi$
           Grade 105, $F_u = 125\,ksi$

Try $1\frac{3}{8}''\,\phi$ Grade 55 Anchor Rod:

   $\phi R_n = 0.75\,(0.75 \times 75)\,\dfrac{\pi (1.375)^2}{4}$

    $\Rightarrow$ $\phi R_n = 62.6\; K/anchor$

Use 2 Anchors to carry T:

    $\phi R_n = 2 \times 62.6 = 125^K > T = 111^K$
                  OK

    Use 2 anchors each side
    ‖ $= 4$, F1554 Grade 55
    ‖ $1\frac{3}{8}''$ Anchor Rods ‖

                   (stiffeners req'd to
                   prevent excessive
                   plate banding)

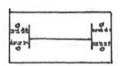

Bibliography

AASHTO. (2011). *Guide Specifications for LRFD Seismic Bridge Design* (2nd ed.). Washington, DC: American Association of State Highway and Transportation Officials.

AASHTO. (2012). *LRFD Bridge Design Specifications* (6th ed.). Washington, DC: American Association of State Highway and Transportation Officials.

AASHTO. (2014). *Guide Specifications for Seismic Isolation Design* (4th ed.). Washington, DC: American Association of State Highway and Tranportation Officials.

AASHTO. (2014). *LRFD Bridge Design Specifications* (7th ed.). Washington, DC: American Association of State Highway and Transportation Officials.

AASHTO. (2017). *AASHTO LRFD Bridge Design Specifications* (8th ed.). Washington, DC: American Association of State Highway and Transportation Officials.

Al Atik, L., & Abrahamson, N. (2010). An Improved Method for Nonstationary Spectral Matching. *Earthquake Spectra, 26*(3), 601–617.

American Concrete Institute. (2013). *ACI 349-13: Code Requirements for Nuclear Safety-Related Concrete Structures and Commentary.* Farmington Hills, MI: ACI.

American Concrete Institute. (2019). *ACI 318-19: Building Code Requirements for Structural Concrete.* Farmington Hills, MI.

American Institute of Steel Construction. (2016). *AISC 341–16: Seismic Provisions for Structural Steel Buildings.* Chicago, IL.

American institute of Steel Construction. (2016). *AISC 358–16: Prequalified Connections for Special and Intermediate Steel Moment Frames for Seismic Applications.* Chicago, IL.

American Institute of Steel Construction. (2016). *AISC 360–16: Specification for Structural Steel Buildings.* Chicago, IL.

American Society of Civil Engineers. (2010). *ASCE/SEI 7–10: Minimum Design Loads for Buildings and Other Structures.* Reston, VA: ASCE/SEI.

American Society of Civil Engineers. (2017). *ASCE 7-16 - Minimum Design Loads and Associated Criteria for Buildings and Other Structures.* Reston, VA: ASCE/SEI.

Applied Technology Council. (2019). *ATC Hazards by Location.* Retrieved April 22, 2019, from https://hazards.atcouncil.org/

ASCE. (2005). *ASCE 43-05: Seismic Design Criteria for Structures, Systems, and Components in Nuclear Facilities.* Reston, VA: American Society of Civil Engineers.

ASCE. (2017). *ASCE 4-16: Seismic Analysis of Safety-Related Nuclear Structures.* Reston, VA: American Society of Civil Engineers.

Atkinson, G., & Beresnev, I. A. (2002). Ground Motions at Memphis and St. Louis from M 7.5–8.0 Earthquakes in the New Madrid Seismic Zone. *Bulletin of the Seismological Society of America, 92*(3), 1015–1024.

Baker, J. W. (2011). The Conditional Mean Spectrum: A Tool for Ground Motion Selection. *Journal of Structural Engineering, 137*(3), 322–331.

Baker, J. W., & Cornell, C. A. (2006). Which Spectral Acceleration Are You Using? *Earthquake Spectra, 22*(2), 293–312.

Baker, J. W., & Jayaram, N. (2008). Correlation of Spectral Acceleration Values from NGA Ground Motion Models. *Earthquake Spectra, 24*(1), 299–317.

Bazzurro, P., & Cornell, C. A. (1999, April). Disaggregation of Seismic Hazard. *Bulletin of the Seismological Society of America, 89*(2), 501–520.

Bentz, E. C. (2000). *Sectional Analysis of Reinforced Concrete* (Ph.D. Thesis). Toronto: University of Toronto.

Bommer, J., Elnashai, A. S., & Weir, A. G. (2000). Compatible Acceleration and Displacement Spectra for Seismic Design Codes. 12th World Conference on Earthquake Engineering (pp. 1–8). Auckland, New Zealand: New Zealand Society for Earthquake Engineering.

Boore, D. M. (2001, October). Effect of Baseline Corrections on Displacements and Response Spectra for Several Recordings of the 1999 Chi-Chi, Taiwan, Earthquake. *Bulletin of the Seismological Society of America, 91*(5), 1199–1211.

Boore, D. M. (2009). *TSPP---A Collection of FORTRAN Programs for Processing and Manipulating Time Series*. Reston, VA: U.S. Geological Survey Open-File Report 2008-1111.

Boore, D. M. (2010, August). Orientation-Independent, Nongeometric-Mean Measures of Seismic Intensity from Two Horizontal Components of Motion. *Bulletin of the Seismological Society of America, 100*(4), 1830–1835.

Boore, D. M. (2020). *TSPP - A Collection of FORTRAN Programs for Processing and Manipulating Time Series*. U.S. Geological Survey Open-File Report 2008-1111.

Boore, D. M., & Akkar, S. (2003). Effect of Causal and Acausal filters on Elastic and Inelastic Response Spectra. *Earthquake Engineering and Structural Dynamics, 32*, 1729–1748.

Boore, D. M., & Bommer, J. J. (2005). Processing of Strong-Motion Accelerograms: Needs, Options and Consequences. *Soil Dynamics and Earthquake Engineering, 25*, 93–115.

Boore, D. M., & Kishida, T. (2017). Relations between Some Horizontal-Component Ground-Motion Intensity Measures Used in Practice. *Bulletin of the Seismological Society of America, 107*(1), 334–343.

Boore, D. M., Watson-Lamprey, J., & Abrahamson, N. A. (2006, August). Orientation-Independent Measures of Ground Motion. *Bulletin of the Seismological Society of America, 96*(4A), 1502–1511.

Bozorgnia, Y., Hachem, M. M., & Campbell, K. W. (2010, February). Deterministic and Probabilistic Predictions of Yield Strength and Inelastic Displacement Spectra. *Earthquake Spectra, 26*(1), 25–40.

Bruneau, M., Uang, C.-M., & Sabelli, R. (2011). *Ductile Design of Steel Structures* (2nd ed.). New York, NY: McGraw-Hill.

Buckle, I. G., Constantinou, M. C., Dicleli, M., & Ghasemi, H. (2006). *MCEER-06-SP07: Seismic Isolation of Highway Bridges*. Buffalo, NY: Multidisciplinary Center for Earthquake Engineering Research.

Buckle, I., Friedland, I., Martin, G., Nutt, R., & Power, M. (2006). *Seismic Retrofitting Manual for Highway Bridges FHWA-HRT-06-032*. Technical Report, Federal Highway Administration, McLean, VA.

Cardone, D., Dolce, M., Matera, F., & Palermo, G. (2008). Application of Direct Displacement Based Design to Multi-Span Simply Supported Deck Bridges with Seismic Isolation: A Case Study. The 14th World Conference on Earthquake Engineering (pp. 1–8). Beijing: International Association for Earthquake Engineering.

Casarotti, C. (2004). *Bridge Isolation and Dissipation Devices*. Pavia, Italy: European School for Advanced Studies in Reduction of Seismic Risk (ROSE School).

Casarotti, C., Pinho, R., & Calvi, G. M. (2005). *Adaptive Pushover-Based Methods for Seismic Assessment and Design of Bridge Structures* (1st ed.). Pavia, Italy: IUSS Press.

Chandler, A. M., Hutchinson, G. L., & Wilson, J. L. (1992). The Use of Interplate Derived Spectra in Intraplate Seismic Regions. 10th World Conference on Earthquake Engineering (pp. 5823–5827). Balkema, Rotterdam: 10th World Conference on Earthquake Engineering.

Charney, F. A. (2010). *NONLIN 8.00 - Computer Program for Nonlinear Dynamic Time History Analysis of Single- and Multi-Degree-of-Freedom Systems*. Blacksburg, VA: Virginia Tech.

Chopra, A. K. (2005). *Earthquake Dynamics of Structures* (2nd ed.). Oakland, CA: Earthquake Engineering Research Institute.

Chopra, A. K. (2016). *Dynamics of Structures* (5th ed.). New York, NY: Pearson.

Christopoulos, C., & Filiatrault, A. (2006). *Principles of Passive Supplemental Damping and Seismic Isolation* (1st ed.). Pavia, Italy: IUSS Press.

Clough, R. W., & Penzien, J. (1975). *Dynamics of Structures*. New York, NY: McGraw-Hill.

Consortium of Organizations for Strong Motion Observation Systems. (2007). COSMOS Virtual Data Center. Retrieved August 26, 2011, from COSMOS Strong Motion Program: http://db.cosmos-eq.org/scripts/default.plx

Dimitriadou, O. (2007, May). *Effect of Isolation on Bridge Seismic Design and Response*. Pavia, Italy: European School for Advanced Studies in Reuction of Seismic Risk (ROSE School).

Duc, T. L. (2007, March). *Verification of the Equations for Equivalent Viscous Damping for Single Degree of Freedom Systems*. Pavia, Italy: European School for Advanced Studies in Reduction of Seismic Risk (ROSE School).

Dynamic Isolations Systems Brochure. (2011). *Seismic Isolation for Buildings and Bridges*. McCarran, NV: Dynamic Isolation Systems.

Earthquake Commission and GNS. (2020, April 18). *Geological hazard information for New Zealand*. Retrieved from GeoNet : https://www.geonet.org.nz/

ESI Group. (2020, April 12). *Scilab*. Retrieved from http://www.scilab.org

Faccioli, E., Paolucci, R., & Rey, J. (2004, May). Displacement Spectra for Long Periods. *Earthquake Spectra, 20*(2), 347–376.

Fardis, M. N., & Pinto, P. E. (2007). *Report No. 2007/05: Guidelines for Displacement-Based Design of Buildings and Bridges*. Pavia, Italy: LESSLOSS - Risk Mitigation for Earthquakes and Landslides.

Federal Emergency Management Agency. (2009). *FEMA P-750: NEHRP Recommended Seismic Provisions for New Buildings and Other Structures: Training and Instructional Materials*. Washington, DC: FEMA.

Fenz, D. M., & Constantinou, M. C. (2008). *Mechanical Behavior of Multi-Spherical Sliding Bearings (MCEER-08-007)*. Buffalo, NY: Multidisciplinary Center for Earthquake Engineering Research.

Fernández, J. A. (2007). *Numerical Simulation of Earthquake Ground Motions in the Upper Mississippi Embayment*. Atlanta, GA: Doctoral Dissertation, Georgia Institute of Technology.

Fernandez, J. A., & Rix, G. J. (2006). Soil Attenuation Relationships and Seismic Hazard Analyses in the Upper Mississippi Embayment. 8th U.S. National Conference on Earthquake Engineering. San Francisco, CA: 8th U.S. National Conference on Earthquake Engineering.

Field, E. H., Jordan, T. H., & Cornell, C. A. (2003, April 20). OpenSHA: A Developing Community - Modeling Environment for Seismic Hazard Analysis. *Seismological Research Letters, 74*(4), 406–419. Retrieved from http://www.opensha.org/

FIP Industriale. (2011). *Lead Rubber Bearings Product Catalog*. Selvazanno, Italy: FIP Industriale.

Frankel, A., Mueller, C., Barnhard, T., Perkins, D., Leyendecker, E. V., Dickman, N., ... Hopper, M. (1996). *National Seismic Hazard Maps: Documentation (OFR 96–532)*. Denver, CO: United States Geological Survey.

Gangopadhyay, A., & Talwani, P. (2003, November/December). Symptomatic Features of Intraplate Earthquakes. *Seismological Research Letters, 74*(6), 863–883.

GeoNET New Zealand. (2012, March 14). Retrieved from http://www.geonet.org.nz/earthquake/

Geschwindner, L. F., Liu, J., & Carter, C. J. (2017). *Unified Design of Steel Structures*. CreateSpace Independent Publishing Platform.

Graizer, V., & Kalkan, E. (2015). *Update of the Graizer-Kalkan Ground-Motion Prediction Equations for Shallow Crustal Continental Earthquakes*. USGS Open-File Report 2015-1009.

Grant, D. N. (2011). Response Spectral Matching of Two Horizontal Ground Motion Components. *Journal of Structural Engineering, 137*(3), 289–297.

Grubb, M. A., Frank, K. H., & Ocel, J. M. (2018). *Bolted Field Splices for Steel Bridge Flexural Members: Overview and Design Examples*. Chicago, IL: National Steel Bridge Alliance.

Gulkan, P., & Sozen, M. A. (1974, December). Inelastic Response of Reinforced Concrete Structures to Earthquake Ground Motions. *Journal of the American Concrete Institute, 71*(12), 604–610.

Hadjian, A. H. (1981). On the Correlation of the Components of Strong Motion - Part 2. *Bulletin of the Seismological Society of America, 71*(4), 1323–1331.

Haghani, R., Al-Emrani, M., & Heshmati, M. (2012). Fatigue Prone Details in Steel Bridges. *Buildings, 2012*(2), 456–476.

Halldorsson, B., & Papageorgiou, A. (2005). Calibration of the Specific Barrier Model to Earthquakes of Different Tectonic Regions. *Bulletin of the Seismological Society of America, 95*(4), 1276–1300.

Hancock, J., Bommer, J. J., & Stafford, P. J. (2008). Number of Scaled and Matched Accelerograms Required for Inelastic Dynamic Analyses. *Earthquake Engineering and Structural Dynamics, 37*(1), 1585–1607.

Harmsen, S. C. (2001). Mean and Modal Epsilon in the Deaggregation of Probabilistic Ground Motion. *Bulletin of the Seismological Society of America, 91*(6), 1537–1552.

Harn, R., Mays, T., & Johnson, G. (2010). *Proposed Seismic Detailing Criteria for Piers and Whatves*. Reston, VA: American Society of Civil Engineers, Ports, 2010.

Hashash, Y. M. (2011). *DEEPSOIL 5: User's Manual and Tutorial*. Urbana-Champaign, IL: UIUC.

Hashash, Y. M., & Park, D. (2001). Non-linear One-dimensional Seismic Ground Motion Propagation in the Mississippi Embayment. *Engineering Geology, 62*, 185–206.

Hashash, Y. M., Tsai, C.-C., Phillips, C., & Park, D. (2008). Soil-Column Depth-Dependent Seismic Site Coefficeints and Hazard Maps for the Upper Mississippi Embayment. *Bulletin of the Seismological Society of America, 98*(4), 2004–2021.

Helwig, T., & Yura, J. (2015). *Steel Bridge Design Handbook - Volume 13: Bracing System Design*. Washington, DC: Federal Highway Administration.

Huang, Y.-N., & Whittaker, A. (2007, August). Scaling Earthquake Records for Response-History Analysis of Safety-Related Nuclear and Conventional Structures. Toronto: Transactions. 19th International Conference on Structural Mechanics in Reactor Technology (SMiRT 19), August 2007, Toronto.

Huff, T. (2016). Partial Isolation as a Design Alternative for Pile Bent Bridges in the New Madrid Seismic Zone. *ASCE Practice Periodical on Structural Design and Construction, 21*(2), 1–12.

Huff, T. (2018). Inelastic Seismic Displacment Amplification for Bridges: Dependence Upon Various Intensity Measures. *ASCE Practice Periodical on Structural Design and Construction, 23*(1), 1–7.

Huff, T. (2020, February). Importance of Target Spectrum Basis in Earthquake Ground Motion Scaling. *ASCE Practice Periodical on Structural Design and Construction, 25*(1), 1–7.

Huff, T., & Pezeshk, S. (2016, May). Inelastic Displacement Spectra for Bridges Using the Substitute-Structure Method. *ASCE Practice Periodical on Structural Design and Construction, 21*(2), 1–13.

Huff, T., & Shoulders, J. (2017). Partial Isolation of a Bridge on Interstate 40 in the New Madrid Seismic Zone. International Bridge Conference. National Harbor, MD, June 2017.

Idriss, I. M., & Boulanger, R. W. (2008). *Soil Liquefaction During Earthquakes* (1st ed.). Oakland, CA: Earthquake Engineering Research Institute.

Iervolina, I., Maddaloni, G., & Cosenza, E. (2008). Eurocode 8 Compliant Real Record Sets for Seismic Analysis of Structures. *Journal of Earthquake Engineering, 12*, 54–90.

Iervolina, I., Maddaloni, G., & Cosenza, E. (2009). A Note on Selection of Time-Histories for Seismic Analysis of Bridges in Eurocode 8. *Journal of Earthquake Engineering, 13*, 1125–1152.

Iervolinoa, I., & Cornell, C. A. (2005, August). Record Selection for Nonlinear Seismic Analysis of Structures. *Earthquake Spectra, 21*(3), 685–713.

IES. (2020, April 12). *Visual Analysis Educational.* Retrieved from http://www.iesweb.com/edu

Integrated Engineering Software, Inc. (2020). *Visual Analysis EDU.* Retrieved January 5, 2020, from https://www.iesweb.com/edu/

Italian Accelerometric Archive. (2010, May). Retrieved December 21, 2011, from Instituto Nazionale di Geofisica e Vulcanologia: http://itaca.mi.ingv.it

Kalkan, E., & Chopra, A. K. (2010). *Practical Guidelines to Select and Scale Earthquake Records for Nonlinear Response History Analysis of Structures: Open File Report 2010–1068.* Reston, VA: U.S. Geological Survey.

Katsanos, E. I., Sextos, A. G., & Manolis, G. D. (2010). Selection of Earthquake Ground Motion Records: A State-of-the-Art Review from a Structural Engineering Perspective. *Soil Dynamics and Earthquake Engineering, 30*(4), 157–169.

Kawashima, K. (2004). Seismic Isolation of Highway Bridges. *Journal of Japan Association for Earthquake Engineering, 4*(3), 283–297.

Kawashima, K., MacRae, G. A., Hoshikuma, J.-i., & Nagaya, K. (1998, May). Residual Displacement Response Spectrum and Its Application. *Journal of Structural Engineering, 124*(5), 523–530.

Kishida, T. (2017, May). Conditional Mean Spectra Given a Vector of Spectral Accelerations at Multiple Periods. *Earthquake Spectra, 33*(2), 469–479.

Kong, C., & Kowalsky, M. J. (2016). Impact of Damping Scaling Factors on Direct Displacement-Based Design. *Earthquake Spectra, 32*(2), 843–859.

Kottke, A., & Rathje, E. (2012). *Technical Manual for SigmaSpectra.* GNU General Public License.

Kottke, A., & Rathje, E. (2020, March 22). *Prof. Ellen M. Rathje - Software.* Retrieved from https://sites.google.com/site/ellenrathje/software-and-data

Kottke, A., & Rathje, E. M. (2008). A Semi-Automated Procedure for Selecting and Scaling Recorded Earthquake Motions for Dynamic Analysis. *Earthquake Spectra, 24*(4), 911–932.

Kunde, M. C., & Jangid, R. (2003). Seismic Behavior of Isolated Bridges: A State-of-the-Art Review. *Electronic Journal of Structural Engineering, 3*, 140–170.

Kwai, T. F. (1986). *Seismic Behavior of Bridges on Isolating Bearings* (Master's Thesis). Christchurch, New Zealand: University of Canterbury.

Lam, N., & Wilson, J. (2004, March). Displacement Modelling of Intraplate Earthquakes. *ISET Journal of Earthquake Technology, 41*(1), 15–52.

Li, Q. L., Liu, M., & Yang, Y. (2002). The 01/26/2001 Bhuj, India, Earthquake: Intraplate or Interplate? *Plate Boundary Zones - AGU Geophysical Monograph*, 255–264.

Liao, W. I., Loh, C. H., & Wan, S. (2000). Responses of Isolated Bridges to Near-Fault Ground Motions Recorded in Chi Chi Earthquake. International Workshop on Annual Commemoration of Chi-Chi Earthquake. Taipei, Taiwan: National Center for Research in Earthquake Engineering.

Lucchini, A., Franchina, P., & Mollaiolia, F. (2017). Spectrum-to-Spectrum Methods for the Generation of Elastic Floor Acceleration Spectra. International Conference on Structural Dynamics, EURODYN 2017 (pp. 3552–3557), Rome, Italy, September 2017.

Luco, N., Ellingwood, B. R., Hamburger, R. O., Hooper, J. D., Kimball, J. K., & Kircher, C. A. (2007). Risk-Targeted versus Current Seismic Design Maps for the Conterminous United States. Structural Engineers Association of California Convention Proceedings. September, 2007, Squaw Creek, California, CA: SEAOC 2007 Convention Proceedings.

Luzi, L., Puglia, R., & Russo, E. (2016). *Engineering Strong Motion Database.* Istituto Nazionale di Geofisica e Vulcanologia, Observatories & Research Facilities for European Seismology. ORFEUS. doi:10.13127/ESM

Lyskova, E. L., Yanovskaya, T. B., & Duda, S. J. (1998). Spectral Characteristics of Earthquakes Along Plate Boundaries. *GEOFIZIKA, 15,* 69–81.

Macrae, G. A., & Kawashima, K. (1997). Post-Earthquake Residual Displacements of Bilinear Oscillators. *Earthquake Engineering and Structural Dynamics, 26,* 701–716.

Madabhushi, G., Knappett, J., & Haigh, S. (2010). *Design of Pile Foundations in Liquefiable Soils* (1st ed.). London: Imperial College Press.

Malaga-Chuquitaype, C., Bommer, J., Pinho, R., & Stafford, P. (2008). Selection and Scaling of Ground Motion Records for Nonlinear Response History Analyses based on Equivalent SDOF Systems. 14th World Conference on Earthquake Engineering. Beijing, October, 2008.

Malekmohammadi, M., & Pezeshk, S. (2014). Nonlinear Site Amplification Factors for Sites Located within the Mississippi Embayment with Consideration for Deep Soil Deposit. *Earthquake Spectra, 31*(2), 699–722.

Maurer Sohne (2011). *Seismic Isolation Systems with Lead Rubber Bearings.* Munich: Maurer Sohne.

McGuire, R. K., Silva, W. J., & Costantino, C. J. (2001). *Technical Basis for Revision of Regulatory Guidance on Design Ground Motions (NUREG/CR-6728).* Washington, DC: U.S. Nuclear Regulatory Commission.

Miller, R., Castrodale, R., Mirmiran, A., & Hastak, M. (n.d.) *NCHRP Report 519: Connection of Simple-Span Precast Concrete Girders for Continuity.* Washington, DC: National Cooperative Highway Research Program.

Mosqueda, G., Whittaker, A. S., Fenves, G. L., & Mahin, S. A. (2004). *Experimental and Analytical Studies of the Friction Pendulum System for the Seismic Protection of Simple Bridges.* Berkeley, CA: Earthquake Engineering Research Center.

Naeim, F., & Lew, M. (1995). On the Use of Design Spectrum Compatible Time Histories. *Earthquake Spectra, 11*(1), 111–127.

Nagarajaiah, S., Reinhorn, A. M., & Constantinou, M. C. (1991). *3D- Basis: Nonlinear Dynamic Analysis of Three-Dimensional Base Isolated Structures: Part II.* Technical Report NCEER-91-0005. Buffalo, NY: National Center for Earthquake Engineering Research, SUNY.

National Research Foundation of Korea. (2020, June 11). *PRISM for Earthquake Engineering.* Retrieved from Earthquake Engineering and Materials Research Group: http://sem .inha.ac.kr/prism/

NEHRP Consultants Joint Venture. (2011). *Selecting and Scaling Earthquake Ground Motions: NIST GCR 11-917-15.* Redwood City, CA: National Institute of Standards and Technology.

Network Center for Earthquake, Tsunami and Volcano, NIED. (2020, April 18). *Strong Motion Seismograph Networks.* Retrieved from https://www.kyoshin.bosai.go.jp/

Olsen, K. B. (2011). *3D Broadband Ground Motion Estimation for Large Earthquakes on the New Madrid Seismic Zone, Central United States - NEHRP Final Report, Award #G10AP00007.* Washington, DC: UCSD: National Earthquake Hazards Reduction Program.

Pacific Earthquake Engineering Research Center. (2020). *PEER NGA-West2 Ground Motion Database*. Retrieved January 6, 2020, from http://ngawest2.berkeley.edu/

Park, D. (2004). *Estimation of Nonlinear Site Effects for Deep Deposits of the Mississippi Embayment* (PhD Thesis). Urbana-Champaign, IL: University of Illinois.

Park, D., & Hashash, Y. M. (2004). Probabilistic Seismic Hazard Analysis with Nonlinear Site Effects in the Mississippi Embayment. 13th World Conference on Earthquake Engineering. Vancouver, BC. Paper No. 1549.

Park, D., & Hashash, Y. M. (2005). Evaluation of Seismic Site Factors in the Mississippi Embayment - I - Estimation of Dynamic Properties. *Soil Dynamics and Earthquake Engineering, 25*, 133–144.

Park, D., & Hashash, Y. M. (2005). Evaluation of Seismic Site Factors in the Mississippi Embayment - II - Probabilistic Seismic Hazard Analysis with Nonlinear Site Effects. *Soil Dynamics and Earthquake Engineering, 25*, 145–156.

Park, D., & Hashash, Y. M. (2004, October). *Estimation of Non-linear Seismic Site Effects for Deep Deposits of the Mississippi Embayment*. Urbana, IL: Mid America Earthquake Center.

Paz, M. (2007). *Structural Dynamics: Theory and Computation*. New York, NY: Springer.

PCI . (2014). *Bridge Design Manual*. Chicago, IL: Precast/Prestressed Concrete Institute.

PEER. (2010, Beta Version, October 1). *Technical Report for the PEER Ground Motion Database Web Application*. Berkeley, CA: University of California, Pacific Earthquake Engineering Research Center.

Petersen, M. D., Frankel, A. D., Harmsen, S. C., Mueller, C. S., Haller, K. M., Wheeler, R. L., ... Rukstales, K. (2008). *Documentation for the 2008 Update of the United States National Seismic Hazard Maps*. Reston, VA: U.S. Geological Survey Open File Report 08-1128.

Petersen, M. D., Moschetti, M. P., Powers, P., Mueller, C., Haller, K., Frankel, A., ... Wheeler, R. (2014). *Documentation for the 2014 Update of the United States National Seismic Hazard Maps*. USGS Open-File Report 2014–1091.

Pezeshk, S., Zandieh, A., & Tavakoli, B. (2011, August). Hybrid Empirical Ground-Motion Prediction Equations fro Eastern North America Using NGA Models and Updated Seismological Parameters. *Bulletin of the Seismological Society of America, 101*(4), 1859–1870.

Pietra, G. M., Calvi, G. M., & Pinho, R. (2008). *Displacement-Based Seismic Design of Isolated Bridges* (1st ed.). Pavia, Italy: IUSS Press.

Priestley, M. J., & Grant, D. N. (2005). Viscous Damping in Seismic Design and Analysis. *Journal of Earthquake Engineering - Imperial College Press, 9*(2), 229–255.

Priestley, M. J., Calvi, G. M., & Kowalsky, M. J. (2007). *Displacement-Based Seismic Design of Structures* (1st ed.). Pavia, Italy: IUSS Press.

Priestley, M. J., Seible, F., & Calvi, G. M. (1996). *Seismic Design and Retrofit of Bridges* (1st ed.). New York: John Wiley & Sons.

Romero, S. M., & Rix, G. J. (2005). *Ground Motion Amplification of Soils in the Upper Mississippi Embayment*. Urbana, IL: NSF/MAE Center.

Ruiz-Garcia, J., & Miranda, E. (2005, August). *Performance-Based Assessment of Existing Structures Accounting for Residual Displacements*. Stanford, CA: The John A. Blume Earthquake Engineering Center, Stanford University.

Ryan, K. L., & Chopra, A. K. (2004, March). Estimation of Seismic Demands on Isolators Based on Nonlinear Analysis. *Journal of Structural Engineering, 130*(3), 392–402.

SeismoSoft. (2020, January). *SeismoArtif 2020*. Retrieved January 5, 2020, from Seismosoft Earthquake Engineering Software Solutions: https://seismosoft.com/products/seismoartif/

SeismoSoft. (2020). *SeismoMatch 2020*. Retrieved January 5, 2020, from Seismosoft Earthquake Engineering Software Solutions: https://seismosoft.com/products/seismomatch/

Seismosoft. (2020). *SeismoSelect 2020*. Retrieved January 5, 2020, from Seismosoft Earthquake Engineering Software Solutions: https://seismosoft.com/products/seismose lect/

Seismosoft. (2020). *SeismoSignal 2020*. Retrieved January 5, 2020, from Seismosoft Earthquake Engineering Software Solutions: https://seismosoft.com/products/seismosi gnal/

SeismoSoft. (2020). *SeismoSpect 2020*. Retrieved January 5, 2020, from Seismosoft Earthquake Engineering Software Solutions: https://seismosoft.com/products/sei smospect

Seismosoft. (2020). *SeismoStruct 2020*. Retrieved January 5, 2020, from Seismosoft Earthquake Engineering Software Solutions: https://seismosoft.com/products/seismost ruct/

Shama, A. A., Mander, J. B., Blabac, B. B., & Chen, S. S. (2001). *Experimental Investigation and Retrofit of Steel Pile Foundations and Pile Bents Under Cyclic Lateral Loadings (MCEER-01-0006)*. Buffalo, NY: Multidisciplinary Center for Earthquake Engineering Research.

Silva, W., Gregor, N., & Darragh, R. (2003). *Development of Regional Hard Rock Attenuation Relations for Central and Eastern North America*. El Cerrito, CA: Pacific Engineering Analysis.

Skinner, R. I., Robinson, W. H., & McVerry, G. H. (1993). *An Introduction to Seismic Isolation* (1st ed.). Chichester, England: Wiley and Sons.

Somerville, P., Collins, N., Abrahamson, N., Graves, R., & Saikia, C. (2001). *Ground Motion Attenuation Relations for the Central and Eastern United States - Final Report*. Pasadena: United States Geological Survey.

Song, S. T., Chai, Y. H., & Hale, T. H. (2004). Limit State Analysis of Fixed-Head Concrete Piles under Lateral Loads. 13th World Conference on Earthquake Engineering. Vancouver, BC: 13th World Conference on Earthquake Engineering, August 2004.

Sritharan, S. (2005). Improved Seismic Design Procedure for Concrete Bridge Joints. *Journal of Structural Engineering, 131*(9), 1334–1344.

Sritharan, S., Fanous, A., Suleiman, M., & Arulmoli, K. (2008). Confinement Reinforcement Requirement for Precast Concrete Piles in High Seismic Regions. 14th World Conference on Earthquake Engineering. Beijing, October 2008.

Stafford, P. J., Mendis, R., & Bommer, J. J. (2008, August). Dependence of Damping Correction Factors for Response Spectra on Duration and Number of Cycles. *Journal of Structural Engineering, 134*(8), 1364–1373.

Stein, S. (2007). Approaches to Continental Intraplate Earthquake Issues. *Continental Intraplate Earthquakes: Science, Hazard, and Policy Issues: Geological Society of America, 425*, 1–16.

Stewart, Jonathan P., Abrahamson, Norman A., Atkinson, Gail M., Baker, Jack W., Boore, David M., Bozorgnia, Yousef, Campbell, Kenneth W., Comartin, Craig D., Idriss, I. M., Lew, Marshall, Mehrain, Michael, Moehle, Jack P., Naeim, Farzad, & Sabol, Thomas A. (2011). Representation of Bidirectional Ground Motions for Design Spectra in Building Codes. *Earthquake Spectra, 27*(3), 927–937.

Stewart, J. P., Chiou, S.-J., Bray, J. D., Graves, R. W., Somerville, P. G., & Abrahamson, N. A. (2001). *Ground Motion Evaluation Procedures for Performance-Based Design*. Berkeley, CA: Pacific Earthquake Engineering Research Center - PEER.

Suarez, V. A. (2008). *Implementation of Direct Displacement Based Design for Highway Bridges*. Raleigh, NC: A dissertation submitted to the Graduate Faculty of North Carolina State University.

Tavakoli, B., & Pezeshk, S. (2005). Empirical-Stochastic Ground-Motion Prediction for Eastern North America. *Bulletin of the Seismological Society of America, 95*(6), 2283–2296.

Toro, G., Abrahamson, N., & Schneider, J. (1997). A Model of Strong Ground Motions from Earthquakes in Central and Eastern North America - Best Estimates and Uncertainties. *Seismological Research Letters, 68*(1), 41–57.

Toro, G. R., & Silva, W. J. (2001). *Scenario Earthquakes for Saint Louis, MO, and Memphis, TN, and Seismic Hazard Maps for the Central United States Region Including the Effect of Site Conditions.* Boulder, CO: USGS.

Towhata, I. (2008). *Geotechnical Earthquake Engineering* (1st ed.). Berlin: Springer-Verlag.

Trombetti, T., Silvestri, S., Gasparini, G., Righi, M., & Ceccoli, C. (2008). Correlations Between the Displacement Spectra and the Parameters Characterizing the Magnitude of the Ground Motion. 14th World Conference on Earthquake Engineering (pp. 1–8). Beijing, October 2008.

United States Geological Survey. (2020). *USGS Unified Hazard Tool.* Retrieved January 6, 2020, from https://earthquake.usgs.gov/hazards/interactive/

USGS and California Geological Survey. (2011). Center for Engineering Strong Motion Data. Retrieved December 21, 2011, from http://strongmotioncenter.org

Vamvatsikos, D., & Cornell, C. A. (2005, August). *Seismic Performance, Capacity, and Reliability of Structures as Seen Through Incremental Dynamic Analysis.* Stanford, CA: The John A. Blume Earthquake Engineering Center, Stanford University.

Van Arsdale, R., & Ellis, M. (2004). *Characterization of Active Faults in the New Madrid Seismic Zone.* Urbana, IL: Mid America Earthquake Center.

Villaverde, R. (2009). *Fundamental Concepts of Earthquake Engineering* (1st ed.). Boca Raton, FL: CRC Press.

Wang, Y.-P., Chung, L.-L., & Liao, W.-H. (1998). Seismic Response Analysis of Bridges Isolated with Friction Pendulum Bearings. *Earthquake Engineering and Structural Dynamics, 27,* 1069–1093.

Warn, G. P. (2002). *Displacement Estimates in Isolated Bridges.* Buffalo, NY: MCEER Student Research Accomplishments.

Warn, G. P., & Whittaker, A. S. (2007). *Performance Estimates for Seismically Isolated Bridges (MCEER-07-0024).* Buffalo, NY: Multidisciplinary Center for Earthquake Engineering Research.

Watson-Lamprey, J. A., & Boore, D. M. (2007, October). Beyond SA-GMRotI: Conversion to SA-Arb, SA-SN, and SA-MaxRot. *Bulletin of the Seismological Society of America, 97*(5), 1511–1524.

Wen, Y. K. (1976). Method for Random Vibration of Hysteretic Systems. *Journal of the Engineering Mechanics Division – ASCE, 102*(2), 249–263.

Wilson, E. L. (1993). *An Efficient Computational Method for the Base Isolation and Energy Dissipation Analysis of Structural Systems - ATC17-1.* Redwood City, CA: Applied Technology Council.

Wong, Y., & Zhao, J. X. (2000). Investigation on Attenuation Characteristics of Strong Ground Motions in China and Hong Kong. 12th World Conference on Earthquake Engineering. Auckland, New Zealand: 12th World Conference on Earthquake Engineering, February 2000.

Wu, C.-L., & Wen, Y. (1999). *Uniform Hazard Ground Motions and Response Spectra for Mid-America Cities.* Urbana, IL: Mid America Earthquake Center.

Xiang, Z., & Li, Y. (2000). Statistical Characteristics of Long Period Response Spectra of Earthquake Ground Motion. 12th World Conference on Earthquake Engineering. Auckland, New Zealand: New Zealand Society for Earthquake Engineering, February 2000.

Index

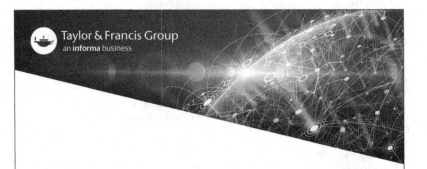

Printed in the United States
By Bookmasters